Residential Design Manual

住宅设计手册

佳图文化 编

2

图书在版编目（CIP）数据

住宅设计手册.2 / 佳图文化编 . — 天津 : 天津大学出版社，2015.1

ISBN 978-7-5618-5203-3

Ⅰ . ①住…… Ⅱ . ①佳… Ⅲ . ①住宅—建筑设计—手册 Ⅳ . ① TU241-62

中国版本图书馆 CIP 数据核字（2014）第 229335 号

责任编辑　油俊伟

出版发行　天津大学出版社
出 版 人　杨欢
地　　址　天津市卫津路 92 号天津大学内（邮编：300072）
电　　话　发行部 022-27403647
网　　址　publish.tju.edu.cn
印　　刷　广州市中天彩色印刷有限公司
经　　销　全国各地新华书店
开　　本　230 mm×300 mm
印　　张　17
字　　数　390 千
版　　次　2015 年 1 月第 1 版
印　　次　2015 年 1 月第 1 次
定　　价　298.00 元

PREFACE 前言

Residence is the most important architectural form of urban landscape. It not only meets people's requirement of living, but also has become an external embodiment of the city's landscape and feature. With the continuous advancement of urbanization and sharply rising urban population, residence is no longer just a shelter in people's view. People now concern more about the beauty of the design and the comfortable sensation of living, and this is a challenging task for most of the residential designers.

Residential Design Manual is a set of professional books that comprehensively introduces and displays the excellent residential design cases at home and abroad. This book specially chooses the highly representative excellent residential designs at home and abroad, a panoramic view with both Chinese and Western style can be seen from the postmodern style residences to townhouses that strive to create an English town style and from waterfront house filled with new Lingnan characteristic to complex buildings with ancient styles. This book is simple but profound; it deeply analyzes every residential design case in the hope of letting the readers feel the beauty of residential design and experience the ingenious thoughts of the designer. Meanwhile, this book is illustrated with abundant exquisite photos in the hope of giving the readers an immersion and letting them feel the beauty essence of the residential design. It is an important reference book for related professionals and those who are practicing and researching on residential design.

住宅场所是城市景观中最重要的建筑形式。它不仅满足了人口居住的需求，而且成为一个城市景观与风貌的外在体现。随着城市化进程的不断推进以及城市人口的急速增多，住宅在人们眼中再也不是单纯的安身之所。人们如今更为看重住宅设计的美观与居住的舒适感。而这也是摆在广大住宅设计师眼前的具有挑战性的课题。

《住宅设计手册》是一套全面介绍和展示国内外优秀住宅设计案例的专业书籍。本书精选国内外极具代表性的优秀住宅设计，从后现代风格住宅到着力营造英国小镇风情的联排别墅，从洋溢新岭南特色的海滨户型到复刻古韵的综合建筑群，中西兼备，尽收眼底。本书深入浅出，对每一个住宅设计案例都进行了深度分析，力求让读者感受到住宅设计之美，体会到设计者心思之巧。同时，本书配以大量精美图片，力求让读者身临其境，置身其中，感受住宅设计的美之真谛。本手册对实践、研究住宅设计的人员以及相关专业人士来说是一套重要的参考书籍。

CONTENTS
目录

New Chinese Style
新中式风格

British Style
英式风格

American Style
美式风格

Italian Style
意大利风格

French Style
法式风格

Other Styles
其他风格

New Chinese Style

新中式风格

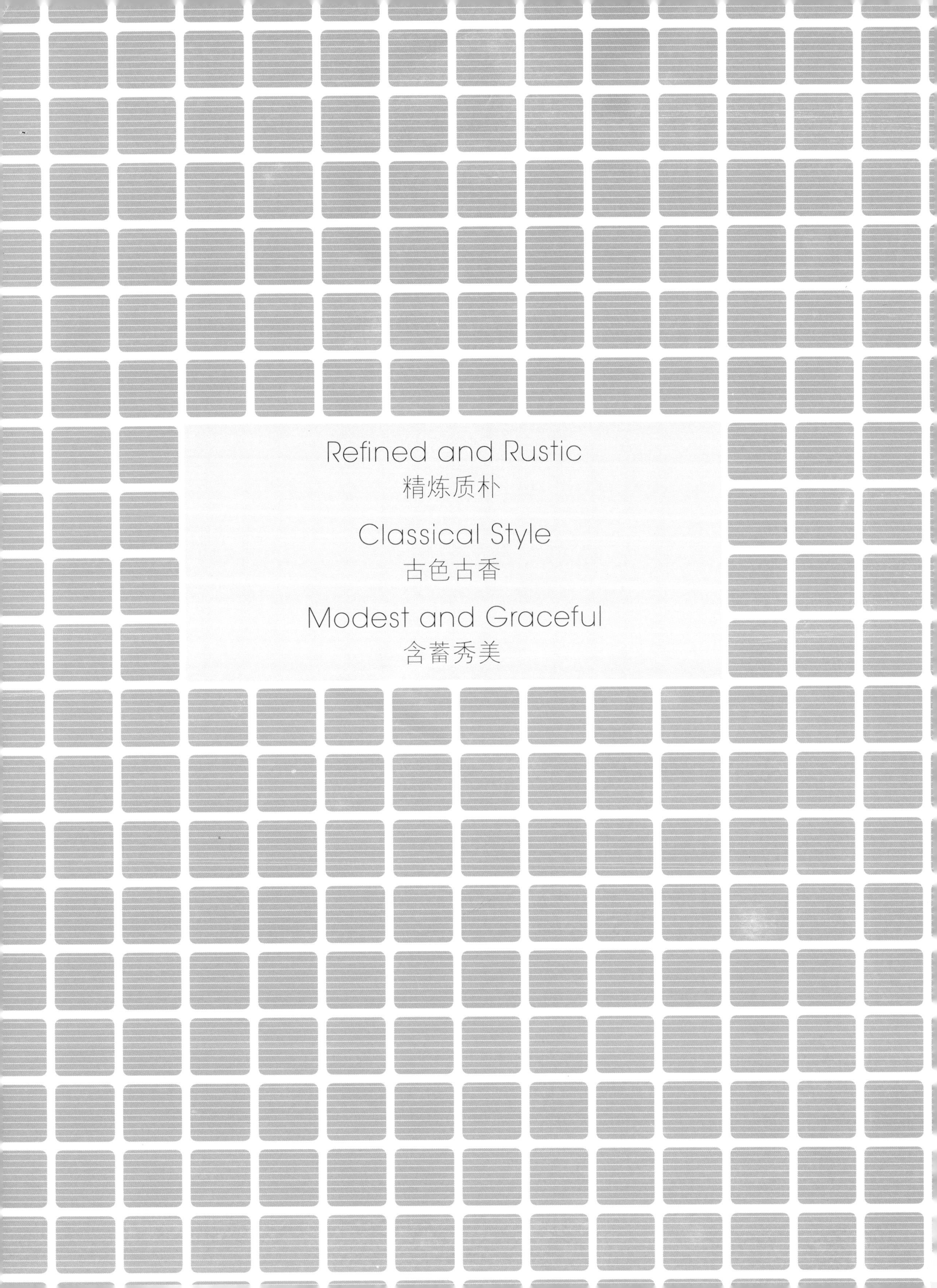

Refined and Rustic

精炼质朴

Classical Style

古色古香

Modest and Graceful

含蓄秀美

Sky Villa
天墅

Location: Xiamen, Fujian, China
Developer: Yongnian Real Estate Development Co., Ltd.
Architectural Design: Tuoruiyimin Architects
Design Team: Huang Yimin, Li Ying, Huang He, Lin Qinghua
Land Area: 13,908.401 m^2
Floor Area: 32,045 m^2
Plot Ratio: 1.8
Green Coverage Ratio: 35%

项目地点：中国福建省厦门市
开发商：永年房地产开发有限公司
设计单位：厦门拓瑞怡民建筑事务所
设计团队：黄奕民　李颖　黄河　林青华
占地面积：13 908.401 m^2
建筑面积：32 045 m^2
容积率：1.8
绿化覆盖率：35%

KEYWORDS 关键词

Foreign Architectural Styles
舶来建筑风格

Xiamen Building
厦门建筑

Local Memory
本土记忆

FEATURES 项目亮点

The exterior design of this project has abandoned the frequently used foreign architectural styles and turned to explore the memory of Xiamen's local buildings, trying to retrieve the local cultural context and interpret the traditional architectural language in a modern way.

项目的整体外观定位摒弃市面上常见的各种舶来建筑风格，转而去发掘厦门建筑的本土记忆，拾回一种属于厦门本土建筑的文化脉络，并对之进行建筑语汇的现代式转译。

➤ Overview

This project is located in the Songbai area, Xiamen, with a gross land area of 13,908.401 m^2 including 9,605.215 m^2 for residential and commercial use in the north, with two 17-storey residential buildings and a 2-storey building for shops there, and 4,303.186 m^2 for a 12-class kindergarten in the south.

➤ 项目概况

项目位于厦门松柏片区，总占地面积13 908.401 m^2，其中商住用地面积9 605.215 m^2，幼儿园用地面积4 303.186 m^2。北侧商住用地由两栋17层的板式高层及沿街两层商业建筑组成，南侧为十二班制幼儿园。

➤ Design Concept

With the concept of taking the advantage of friendly environment and creating a product different from the conventional concrete residential box, the architects want to provide residents with more outdoor spaces and communication places, to make them return to the leisurely and cozy life of living in a mansion with the sense of old Xiamen.

➤ 设计理念

整体规划理念希望依托区位宜居的环境，打造一种与常规的钢筋水泥式住宅盒子不同的产品，让人有更多的户外生活空间、更多的交流场所，回归一种悠闲、慵懒，带有老厦门味道的别墅式大宅生活。

永年地产
永年 天玺

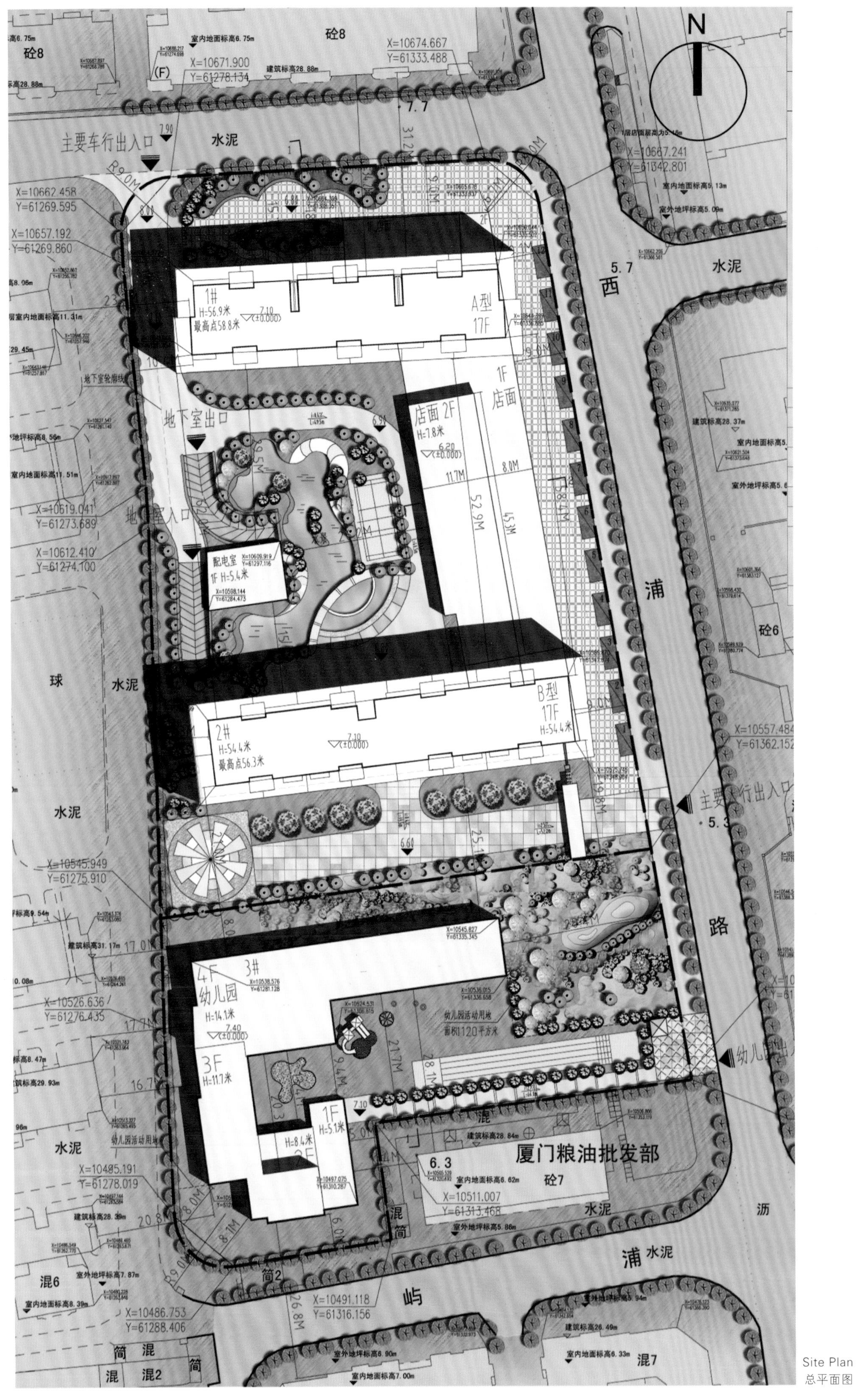

Site Plan
总平面图

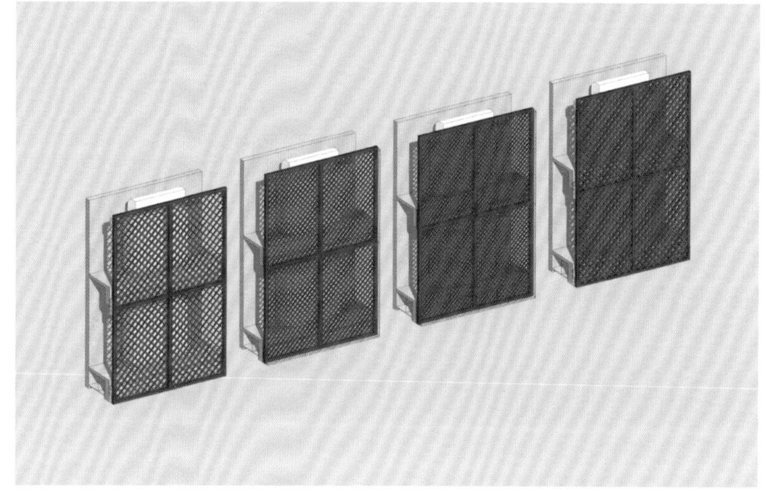

➤ Overall Planning

From the view of overall plane layout, in order to make full use of the depth of the land, two 17-storey residential buildings are arranged in parallel in the north and south. A 50 m x 55 m green atrium is created in the dense and old urban area, and a 2-storey building for shops is set in the east to cut off the traffic noise. A garden entry courtyard with the size of 20 m x 70 m is designed in the south to connect the residential buildings and the kindergarten. In this way, it has maximized the public green space and ensured that each unit can enjoy the view of private hanging garden, achieving an excellent landscape design that small gardens are in the large courtyard, gardens are only a few steps away from the residences, and privacy and openness are cooperating well. In the Sky Villa, residents can sit in their own private hanging gardens to listen to the birds and enjoy the sunshine that the courtyard brings to them.

➤ 总体规划

从总平面布局上，两栋17层的空中别墅住宅南北平行布置，最大限度地利用了用地的纵深。在密集的老市区开辟出50 m×55 m的绿色中庭，东侧两层商业店面的设置进一步隔绝了外面道路的喧嚣，南侧在与幼儿园交接的位置，布置了20 m×70 m的花园庭院作为入口。在此布局下，社区公共绿化空间实现最大化，住宅布局使得户户南北通透，并且每一户都拥有私家空中花园；社区景观上做到园中有园、自然渗透入户、私密与公共良好过渡的绝佳状态。在天墅，住户可以在自己私密的空中花园里，享受闹市中的绿色庭院带来的鸟叫蝉鸣与阳光空气。

➤ Facade Design

The exterior design of this project has abandoned the frequently used foreign architectural styles and turned to explore the memory of Xiamen's local buildings, trying to retrieve the local cultural context and interpret the traditional architectural language in a modern way. After the exploration, the architects are eager to find a way to combine the tradition with modernity to redefine the classics.

The image design, from the cornice, balcony and garden space, to the window-wall ratio, texture and grid and entrance, is inspired by the memories of the mansions of old Xiamen. In cooperation with modern construction and the requirements of project budget, the architects ignore the conventional architectural styles and explain the characters of Xiamen in this project.

➤ 立面设计

本项目的整体外观定位摒弃市面上常见的各种舶来建筑风格，转而去发掘厦门建筑的本土记忆，拾回一种属于厦门本土建筑的文化脉络，并对之进行建筑语汇的现代式转译。希望通过设计师的探索，找到传统结合现代的方式，重新诠释经典。

天墅的形象设计，无论是建筑的檐口、阳台样式、花园空间，还是墙面的开窗比例、网格肌理、入户节奏，都来自厦门旧时大宅的某些回忆片段，并结合当代的施工造诣和工程预算要求，在各种主义充斥市场的现状下，将这份厦门血统延续在天墅的设计当中。

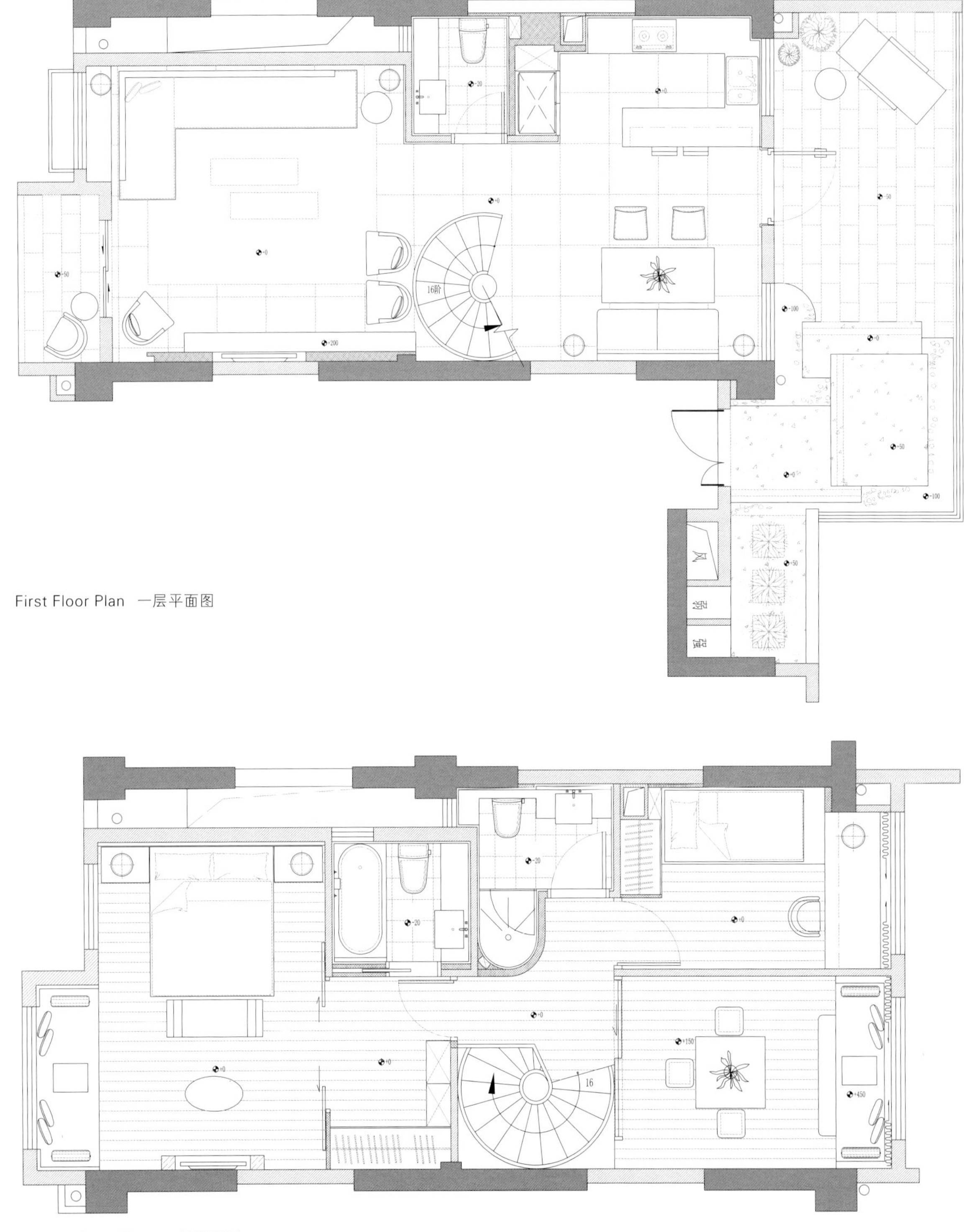

First Floor Plan 一层平面图

Second Floor Plan 二层平面图

➤ House Type Design

All the units are designed as duplex apartments, and staggered unit entrance layout has ensured each floor with only one unit entrance, but also provided residents with privacy and prestige. 4-storey high and 3 m wide private hanging gardens are created so that all the residents living in this high density city can enjoy a garden life.

➤ 户型设计

户型为一梯两户的单元式复式住宅，每单元两户之间入户层错开布置，让电梯系统每层只服务于一户，做到绝佳私密性和别墅式独门入户的尊贵感。在这里，每一户都拥有一个挑高4层、临空出挑 3 m的私人空中花园。设计师试图为每一位生活在高密度城市空间中的居住者带去花园式的生活场景。

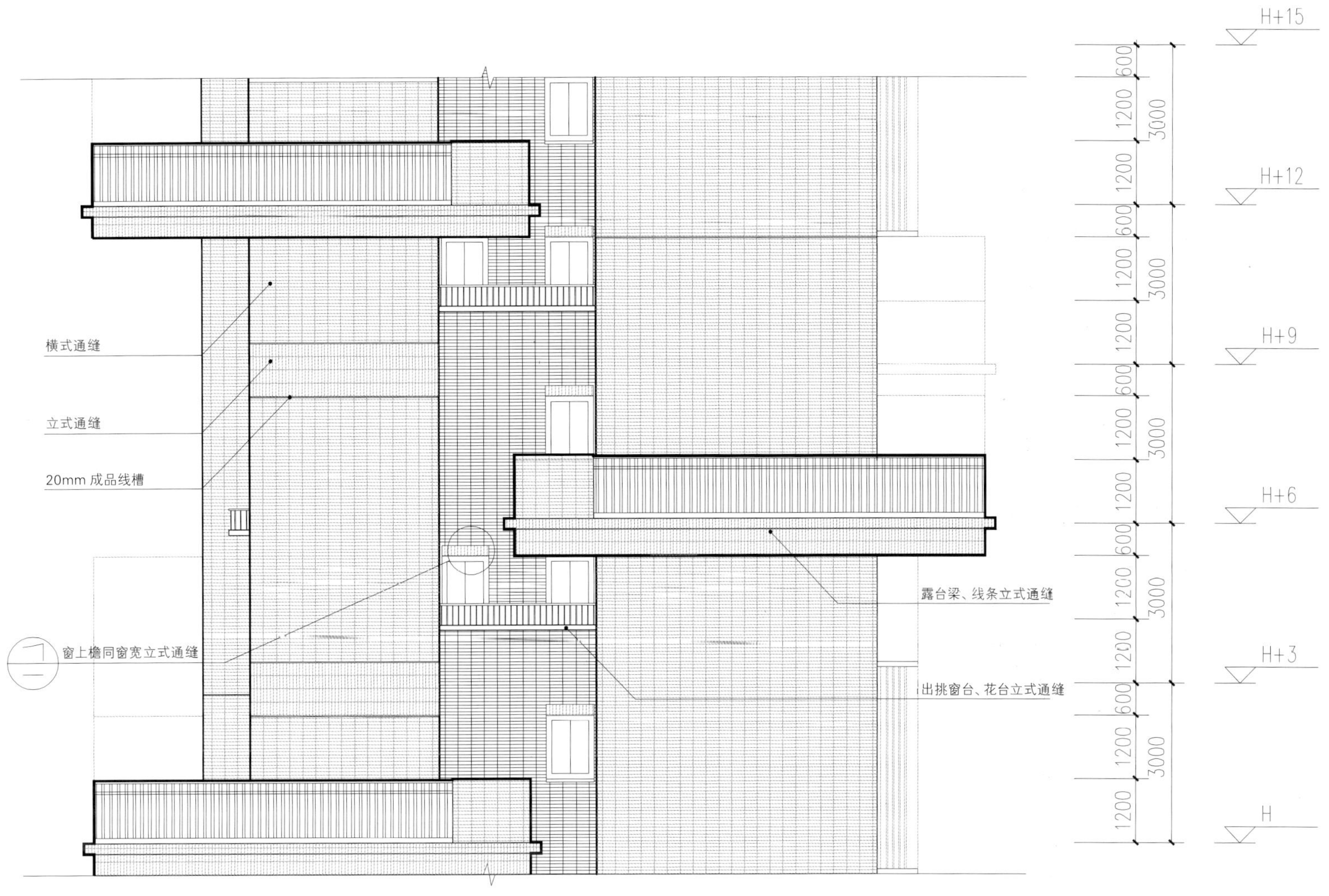

Elevation 1 立面图 1

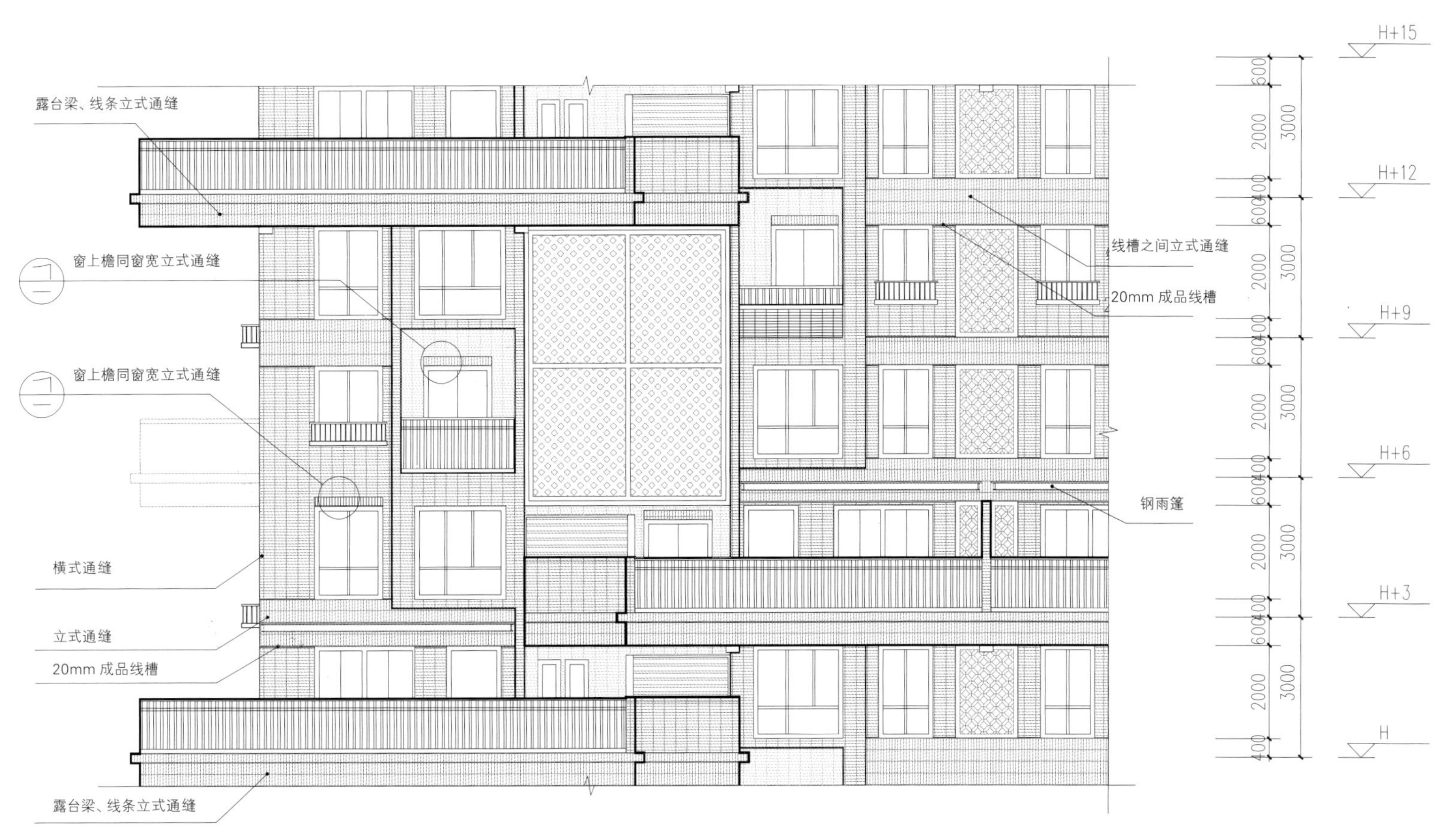

Elevation 2 立面图 2

Nanjing CNK • Tangshan Residence

南京城开 · 汤山公馆

Location: Nanjing, Jiangsu, China
Developer: Nanjing CNK Real Estate Group
Architectural Design: Nanjing Yangtze River Urban Architectural Design Co., Ltd.
Shanghai Peddle Thorp Architecture Design & Consultant Co., Ltd.

项目地点：中国江苏省南京市
开发商：南京城开房地产置业有限责任公司
设计单位：南京长江都市建筑设计股份有限公司
上海柏涛建筑设计咨询有限公司

FEATURES 项目亮点

Nanjing CNK · Tangshan Residence perfectly integrates the architectural style of the Republic of China with modern and fashionable elements, and introduces hot spring to each unit. With the smallest apartment measuring about 60 m^2, it will be the second residence for the working class.

南京城开 · 汤山公馆项目采用民国建筑风格，并结合了现代时尚元素。建成后，温泉将被引入每家每户。该项目的公寓形态中，最小面积约为 60 m^2，项目定位为工薪阶层的第二居所。

KEYWORDS 关键词

Architectural Style of the Republic of China
民国建筑风格

Modern and Stylish
现代时尚感

Working Class Residence
工薪阶层住房

Overview

The project, with low density and multi-stories, is located in Tangshan Town, eastern suburb of Nanjing, 1 kilometer away from the Tangshan exit of Shanghai-Ningbo Expressway, to the south of Wenquan Road, west of Tangshui River and north of Tangtong Road.

项目概况

项目为低密度多层住宅，位于南京市东郊汤山镇，具体位置为沪宁高速汤山出口1 km处，温泉路以南，汤水河以西，汤铜路以北。

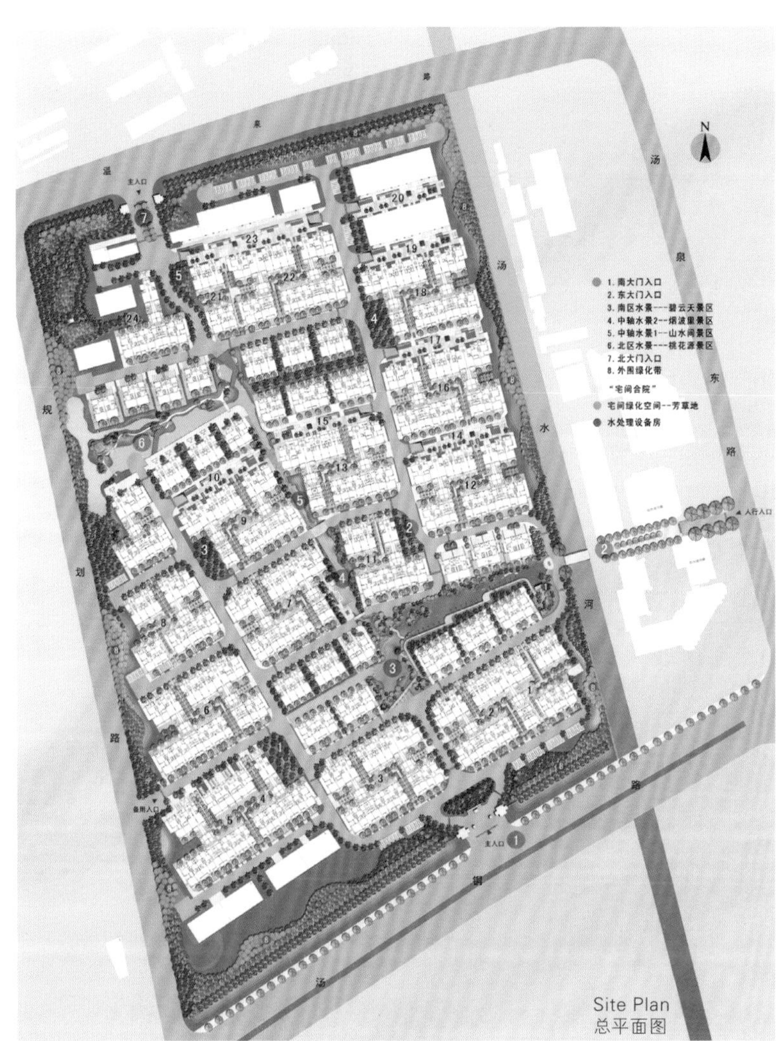

Site Plan
总平面图

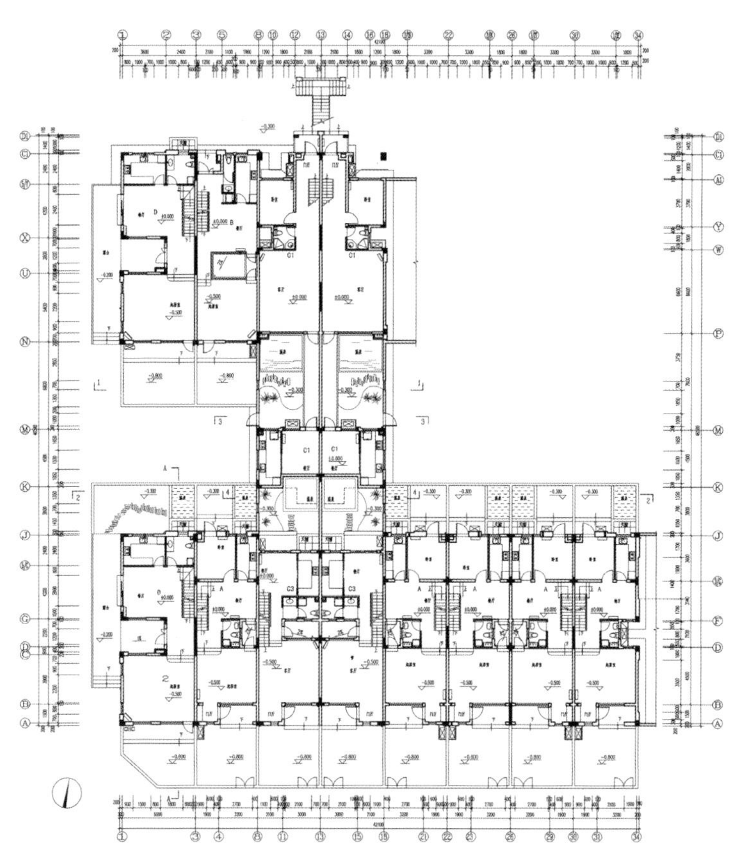

Building 02-1 First Floor Plan
02-1 幢一层平面图

Project Orientation

The project perfectly integrates the architectural style of the Republic of China with modern and fashionable elements, and introduces hot spring to each unit. With the smallest apartment measuring about 60 m^2, it will be the second residence for the working class.

Following the masterpieces of Jinling Mingzuo, Jinling Mingrenju, CNK Yijia, Beiyuan Star and CNK Building International, this project is another masterpiece of CNK Group with architectural style of the Republic of China. It is a high-end community gathering hot spring, landscape, alleys and the sentiment of the Republic of China, which is a perfect choice for those who are chasing after the high quality life.

项目定位

项目采用民国建筑风格，并结合了现代时尚元素。建成后，温泉将被引入每家每户。该项目的公寓形态中，最小面积约为60 m^2，项目定位为工薪阶层的第二居所。

该项目是南京城开集团继金陵名座、金陵名人居、城开怡家、北苑之星、城开建筑国际等一批成功楼盘之后的又一力作，也是南京城开集团倾心打造的民国风格住宅，是集温泉、山水、街巷、民国情调于一体的高档生活区，是追求高品质生活人士的完美居所。

➤ Project Types

With a total unit number of 715, it has a land area of 145,000 m², gross floor area of 122,000 m² and aboveground floor area of 106,000 m², including 104,000 m² for residence and 2,000 m² for public area. The plot ratio is 0.73 and the highest height of the building is 16 m. There are 2-storey semidetached villas, 3-storey townhouses, 4-storey superimposed villas, 3- to 4-storey courtyard villas and 4- to 5-storey lift apartments in the project.

➤ 项目类型

住宅小区总户数715户；占地面积145 000 m²，总建筑面积122 000 m²，地上总建筑面积106 000 m²，其中住宅面积104 000 m²，公建面积2 000 m²，容积率0.73，建筑最高达到16 m。住宅类型分别为2层的双拼别墅、3层的联排别墅、4层的叠加别墅、3～4层的合院别墅和4～5层的电梯公寓。

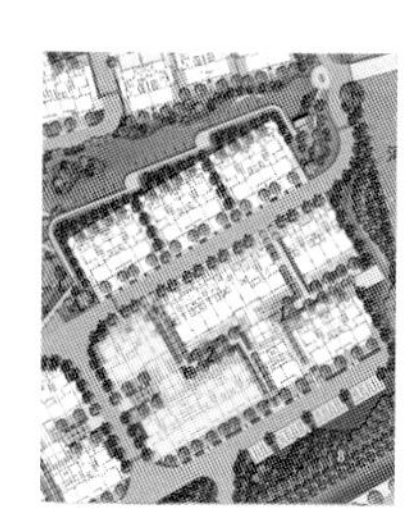

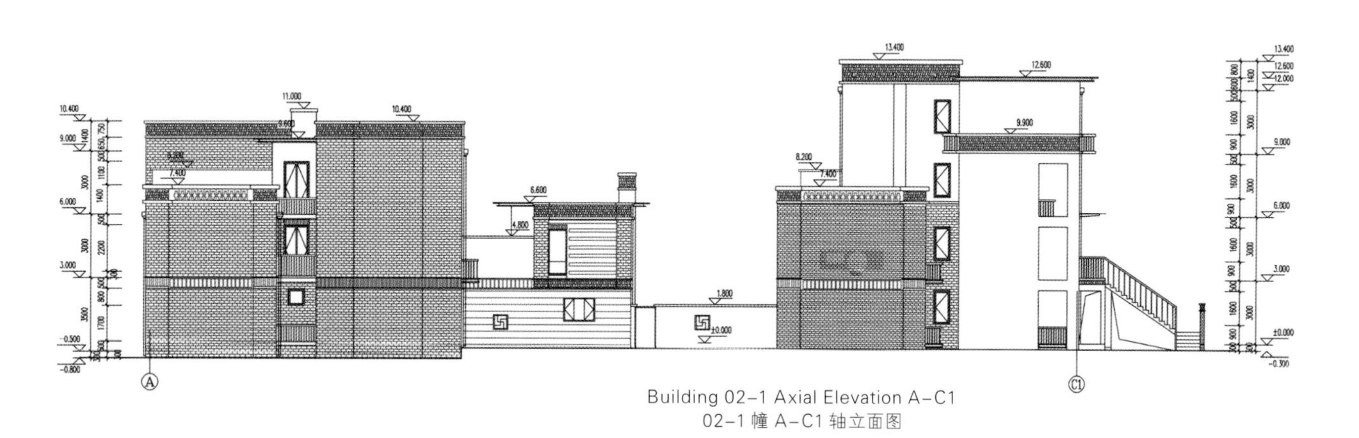

Building 02–1 Axial Elevation A–C1
02–1 幢 A–C1 轴立面图

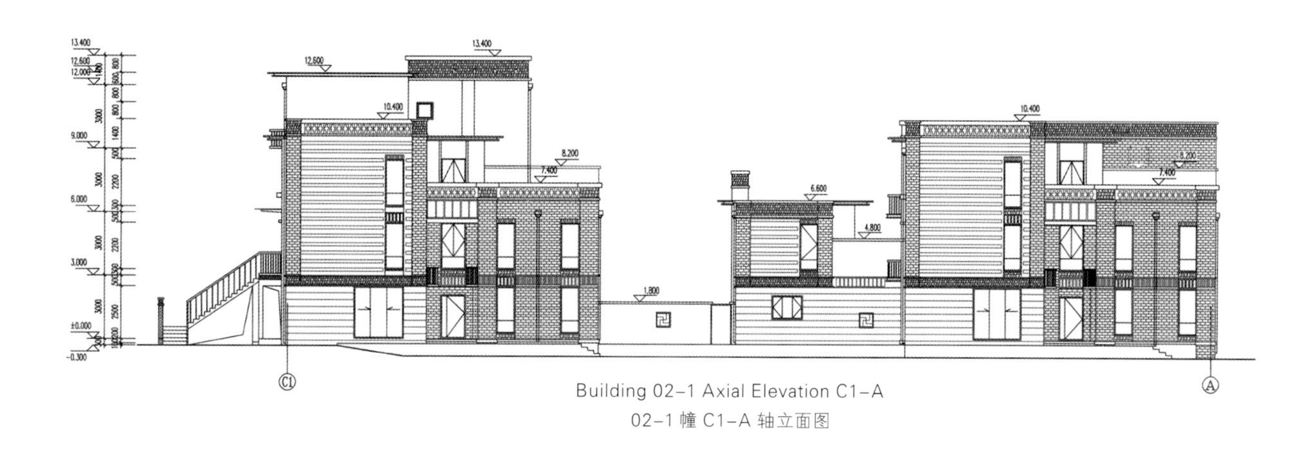

Building 02–1 Axial Elevation C1–A
02–1 幢 C1–A 轴立面图

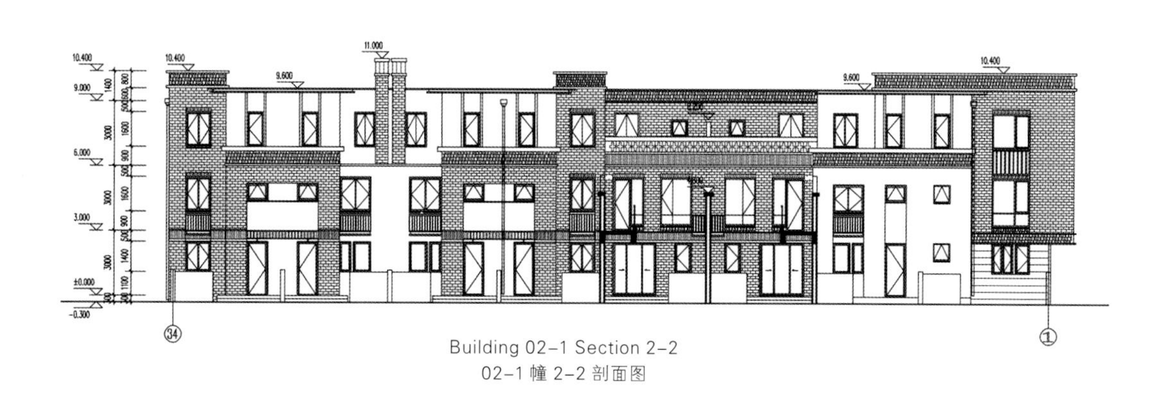

Building 02–1 Section 2–2
02–1 幢 2–2 剖面图

Building 02–1 Section 1–1
02–1 幢 1–1 剖面图

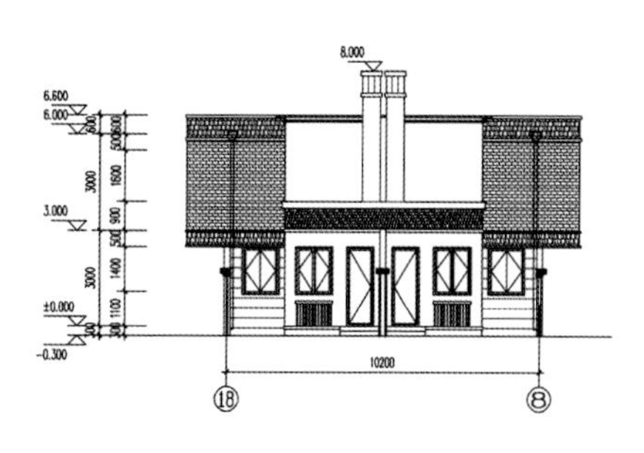

Building 02–1 Section 3–3
02–1 幢 3–3 剖面图

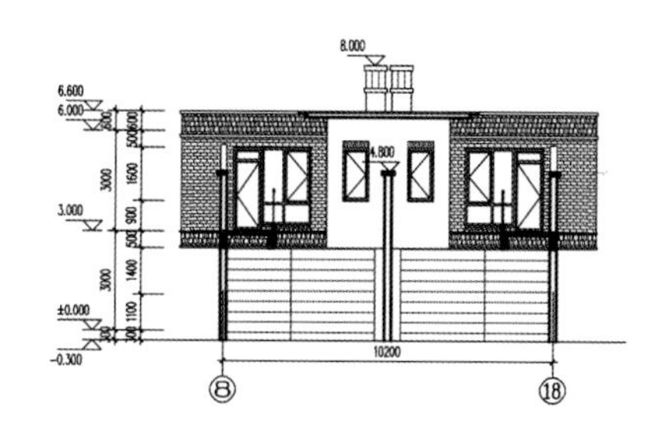

Building 02–1 Section 4–4
02–1 幢 4–4 剖面图

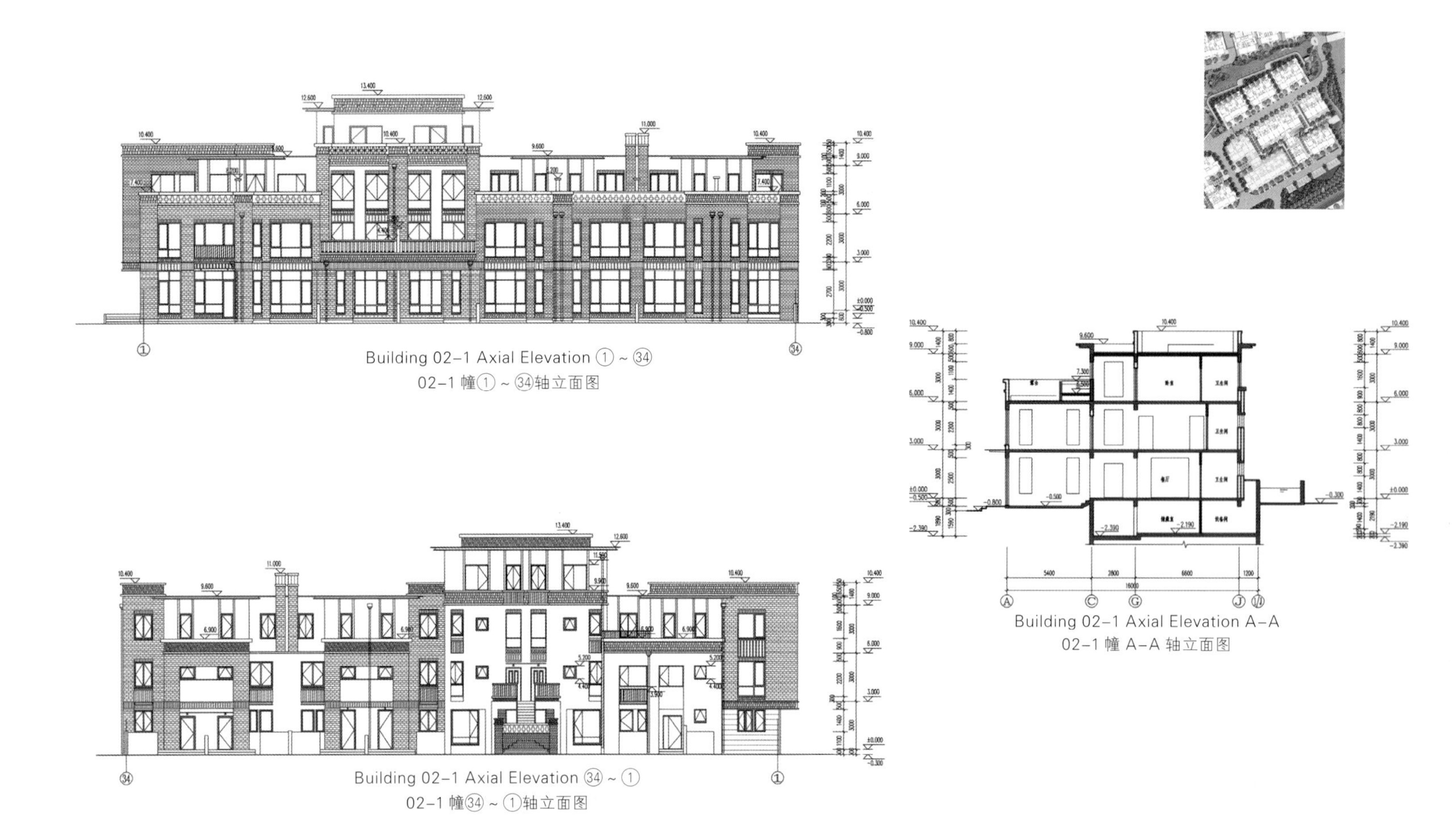

Building 02-1 Axial Elevation ① ~ ㉞

02-1 幢① ~ ㉞轴立面图

Building 02-1 Axial Elevation A-A

02-1 幢 A-A 轴立面图

Building 02-1 Axial Elevation ㉞ ~ ①

02-1 幢㉞ ~ ①轴立面图

Yadong • Natural Garden (Phase I)
亚东·朴园（一期）

Location: Zhenjiang, Jiangsu, China
Architectural Design: Nanjing Yangtze River Urban Architectural Design Co., Ltd.
Overground Floor Area: 29,536 m^2
Underground Floor Area: 8,588 m^2
Plot Ratio: 0.72
Building Density: 23.73%
Green Coverage Ratio: 44.38%

项目地点：中国江苏省镇江市
设计单位：南京长江都市建筑设计股份有限公司
地上总建筑面积：29 536 m^2
地下总建筑面积：8 588 m^2
容积率：0.72
建筑密度：23.73%
绿化覆盖率：44.38%

FEATURES 项目亮点

The residences are designed with frontyards, patios and backyards, integrating plants and landscape, to create a unique Chinese residential cultural conception and reshape the courtyard lifestyle that belongs to the Chinese.

住宅设计了前庭、天井和后院，将大量的植物景观融入宅院本身，形成独特的中国居住文化意境，重塑一种真正属于中国人的庭院生活方式。

KEYWORDS 关键词

Plants and Landscape
植物景观

Cultural Conception
文化意境

Courtyard Lifestyle
庭院生活方式

Overview

With a gross land area of 40,000 m^2, this project has a club and 19 residential buildings which are townhouses and superimposed houses, with 152 units in total.

项目概况

亚东·朴园一期基地总占地面积为40 000 m^2，工程包括会所及19栋住宅，住宅为联排住宅与叠加住宅，住户为152户。

Planning Layout

The site is in the open space between the north and south, and serves as the viewing corridor for the mountains around. The planning structure can be explained by a loop and an axis: the loop is the primary traffic as well as the landscape ring, while the axis mainly composed of water body has expressed the feature of this mountain view residence. The west is for superimposed houses, and the east is for townhouses, with landscape along the roads between them. The public facilities, which are for community services, cultural activities and a few small businesses, are placed on the main entrance on the north, so that they are able to provide services for the planned plot 5.

规划布局

整个地块以南北共享的开放空间主轴来串联，同时作为对外部山体的景观视廊。规划结构可用一环一轴来表示：一环即主要的交通环，同时也是绿环；一轴即以中心水体为主体的开放空间轴，以水的灵魂来点出山景住宅的特色。具体布局为：西区叠加、东区联排。联排与叠加之间沿道路设置绿化景观带。公建布置在小区北侧主入口处，以便其服务半径可辐射到后期开发的5号地块，功能主要为社区服务、文化活动站及部分小型商业。

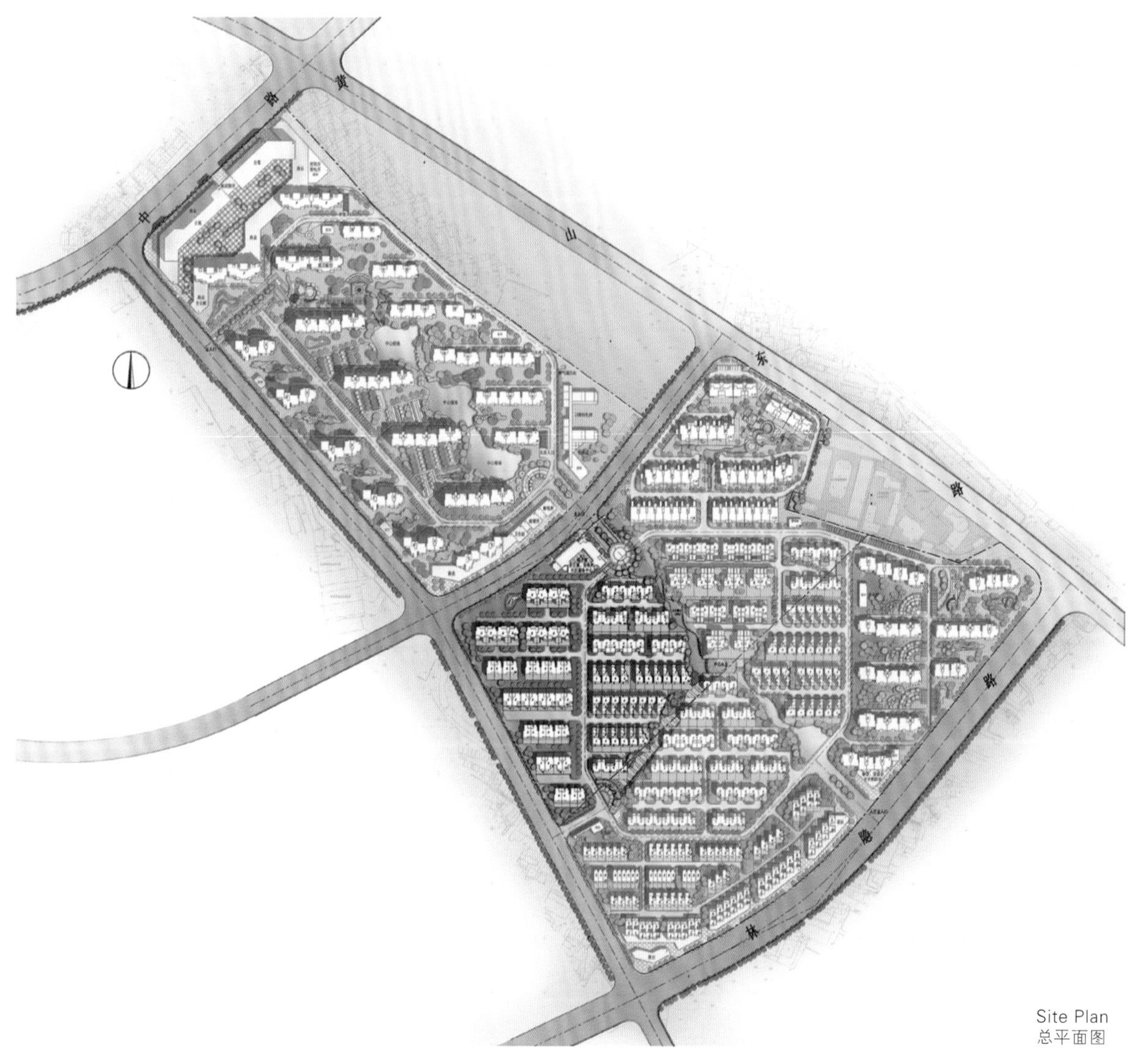

Site Plan
总平面图

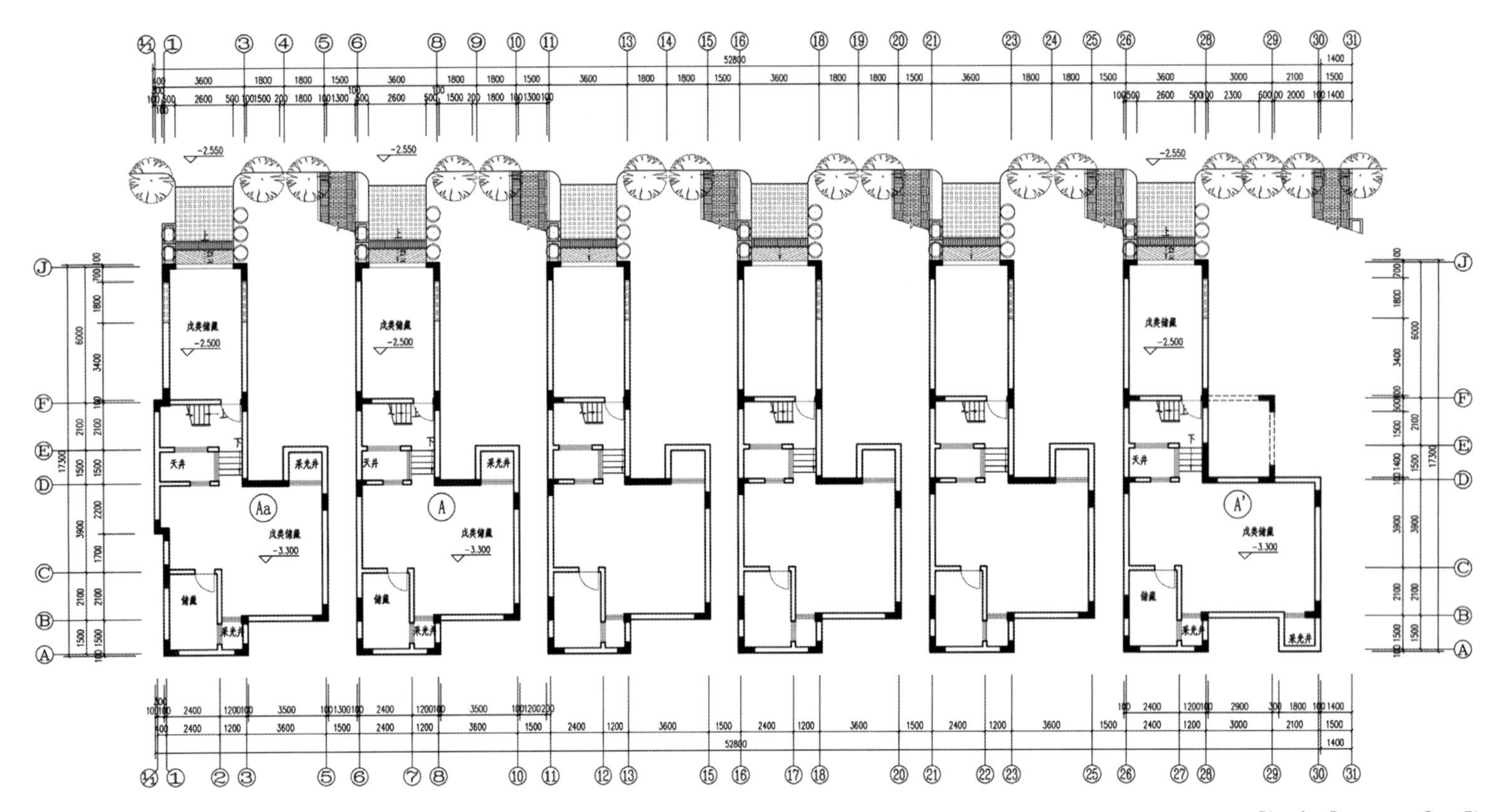

Plan for Basement One Floor
地下一层平面图

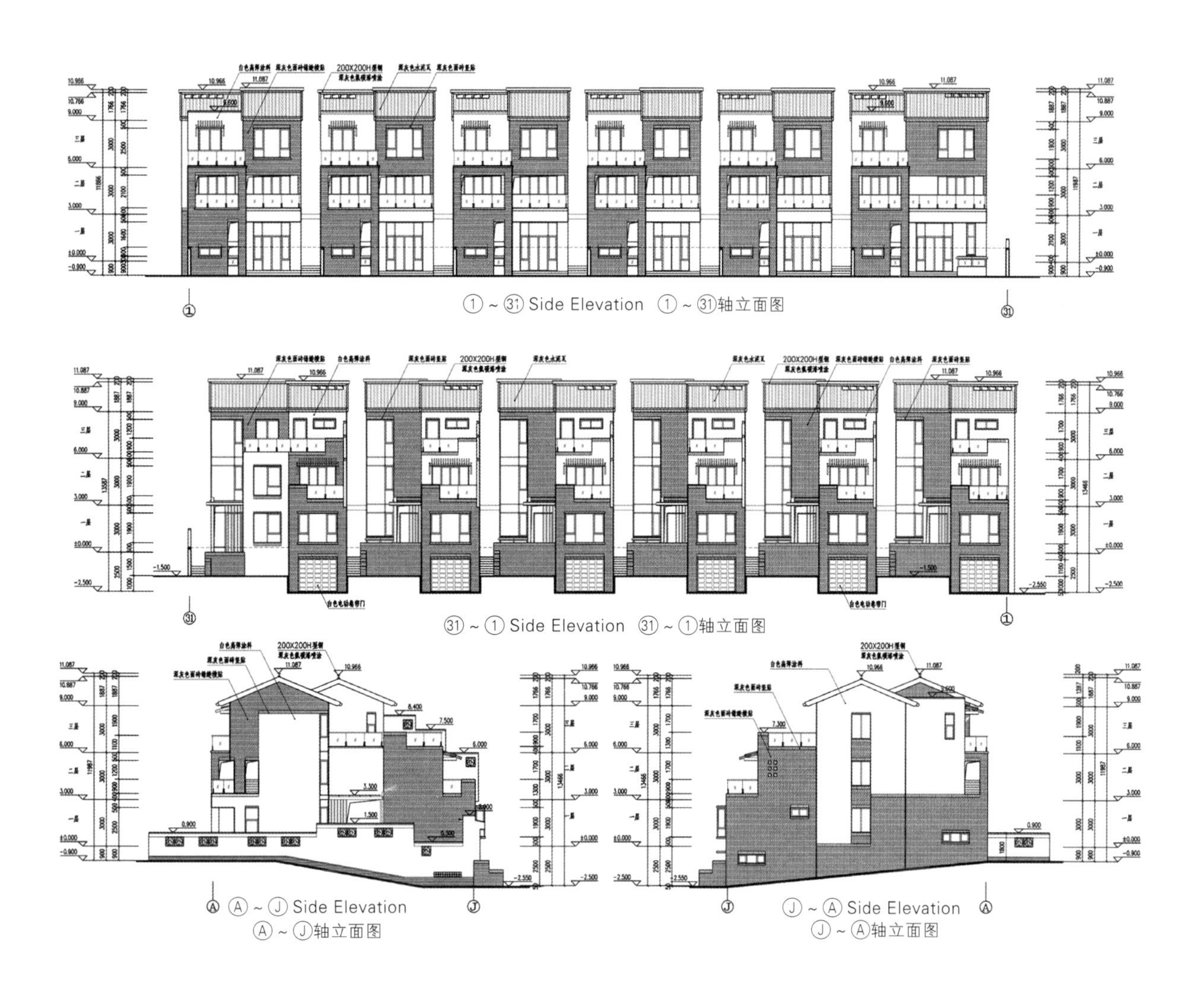

① ~ ㉛ Side Elevation ① ~ ㉛轴立面图

㉛ ~ ① Side Elevation ㉛ ~ ①轴立面图

Ⓐ ~ Ⓙ Side Elevation
Ⓐ ~ Ⓙ轴立面图

Ⓙ ~ Ⓐ Side Elevation
Ⓙ ~ Ⓐ轴立面图

➤ Architectural Design

Combining the cultural deposits of Zhenjiang with traditional Chinese residential value, the architects have considered the courtyard as the core element. The residences are designed with frontyards, patios and backyards, integrating plants and landscape, to create a unique Chinese residential cultural conception and reshape the courtyard lifestyle that belongs to the Chinese.

➤ 建筑设计

项目将镇江的文化底蕴与中国传统居住价值相结合，把庭院作为核心元素来考虑。住宅设计了前庭、天井和后院，将大量的植物景观融入宅院本身，形成独特的中国居住文化意境，重塑一种真正属于中国人的庭院生活方式。

1-1 Section
1-1剖面图
2-2 Section
2-2剖面图
3-3 Section
3-3 剖面图
4-4 Section
4-4剖面图
5-5 Section
5-5 剖面图
6-6 Section
6-6 剖面图

Zhongdu Wonderland

中都青山湖畔 · 绿野清风

Location: Lin'an, Zhejiang, China
Architectural Design: The Architectural Design & Research Institute of Zhejiang University Co., Ltd.
Gross Site Area: 144,000 m^2
Gross Floor Area: 39,750 m^2
Gross Land Area: 23,240 m^2
Commercial Floor Area: 7,950 m^2
Club Floor Area: 3,570 m^2
Villa Floor Area: 28,230 m^2
Plot Ratio: 0.276
Building Density: 16.1%
Green Coverage Ratio: 62.4%

项目地点：中国浙江省临安市
设计单位：浙江大学建筑设计研究院
总用地面积：144 000 m^2
总建筑面积：39 750 m^2
总占地面积：23 240 m^2
商业建筑面积：7 950 m^2
会馆建筑面积：3 570 m^2
别墅建筑面积：28 230 m^2
容积率：0.276
建筑密度：16.1%
绿化覆盖率：62.4%

KEYWORDS 关键词

Courtyard Space
院落空间

Dramatic Structure
戏剧化形体

Building Along Slope
顺坡而建

FEATURES 项目亮点

By integrating and borrowing the elements from traditional Chinese architectural culture, and based on the topographical situation, the architects have designed lots of open and semi-open courtyard spaces and grey spaces, which express the charm of traditional Chinese courtyard spaces.

设计融合和借鉴了许多中国传统建筑文化的设计要素，依据地形设计了许多开敞与半开敞的院落空间和灰空间，每个角落都能体现出中国传统围合院落空间的魅力。

➤ Overview

This project is located in the southwest of Qingshan Lake of Lin'an, and west to Lin'an urban area. The landform is mainly composed of slopes with 30–40 degree to the max. It is mainly with mono-residences and supplemented with commercial buildings and supporting commercial buildings.

➤ 项目概况

项目位于临安市著名的青山湖旅游风景区的西南角，西侧与临安市区相邻。地貌以坡地地形为主，坡度最大达到30~40度。规划主要建筑以山地单体住宅为主，同时还包括营业用房与附属配套商业用房等。

➤ Architectural Design

By integrating and borrowing the elements from traditional Chinese architectural culture, and based on the topographical situation, the architects have designed lots of open and semi-open courtyard spaces and grey spaces, which express the charm of traditional Chinese courtyard spaces. For the architectural form, the architects use simple facade materials, rich space design and dramatic structure to achieve the combination of mountains and buildings. The residences are built along the slopes, so that they are not only echoing with the slopes in the structure, but also decreasing the cut & fill earthwork volume and the effects on the mountains and nature. Before the architectural design, a set of measurements and setting-out-of-route program has been conducted, to avoid the effects on the plants.

➤ 建设设计

设计融合和借鉴了许多中国传统建筑文化的设计要素，依据地形设计了许多开敞与半开敞的院落空间和灰空间，每个角落都能体现出中国传统围合院落空间的魅力。在建筑形态上，通过简洁的立面语言和材料的应用、丰富的空间设计、戏剧化的形体处理来表现建筑和山地的融合。建筑顺坡而建，不仅从形式上能够和山坡相呼应，而且可以减少填挖方，减少对山体和自然的破坏。设计前对树木进行测绘放线，尽量避免移动大树，建造时有计划地保留部分原生植被。

主要技术经济指标

项目		计算单位	数值
总用地面积		m²	144000
总建筑面积		m²	39750
其中：	商业建筑面积	m²	7950
	会馆建筑面积	m²	3570
	别墅建筑面积	m²	28230
建筑总占地面积		m²	23240
其中：	商业占地面积	m²	4000
	会馆占地面积	m²	1600
	别墅建筑占地面积	m²	17640
容积率			0.276
建筑密度		%	16.1
绿地率		%	62.4
别墅总户数		户	97
停车位		辆	204

Site Plan
总平面图

➤ Architectural Details and Humanization

By making use of the characters of the mountain land, lots of flexible spaces are created to make the residences richer and more diverse in the plane design. Furthermore, indoor barrier-free elevators are set to create an easier life.

➤ 建筑细部与人性化

平面利用山地特点合理设计了很多可以灵活使用的空间，使住宅室内具有更丰富和多样性的效果。在山地住宅内设计了户内无障碍电梯，方便住户的起居生活。

➤ Green and Environmental-Friendly

Lots of innovative and green and environmental-friendly building materials, such as section steel and woods, as well as low-technology and low-investment green and energy-saving designs, are used for the project. Learning from the characters of caves that absorb the warmth of the mountains, the buildings are integrated with the mountains and nature to ensure a cool summer and warm winter.

➤ 绿色环保

项目创新性地使用了大量型钢、木材等绿色环保建筑材料，同时在设计中使用了低技术和低投入的绿色与节能手法。借鉴传统窑洞的优点，充分利用山体特点，居住在里面可以明显感受到冬暖夏凉的自然温度调节。

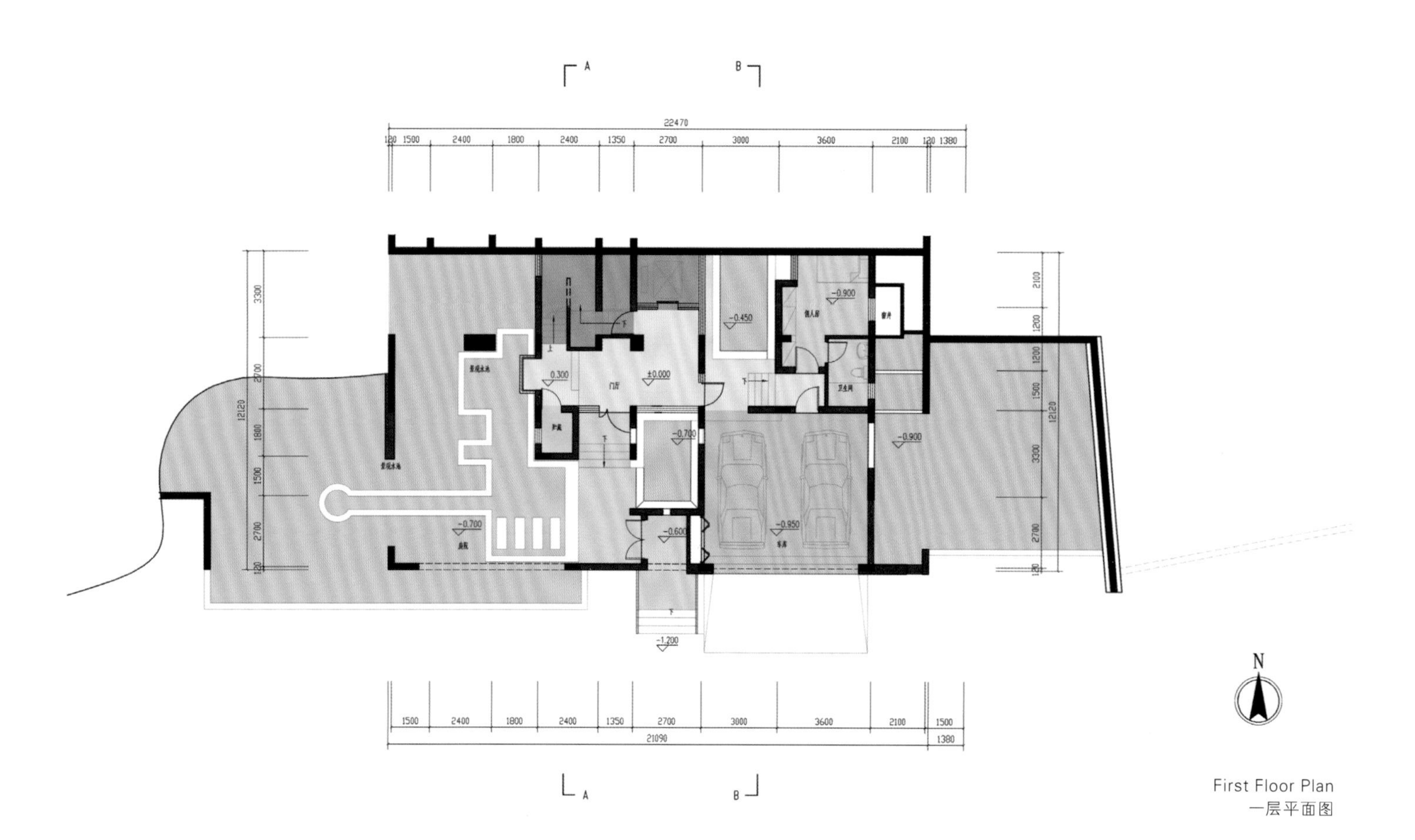

First Floor Plan
一层平面图

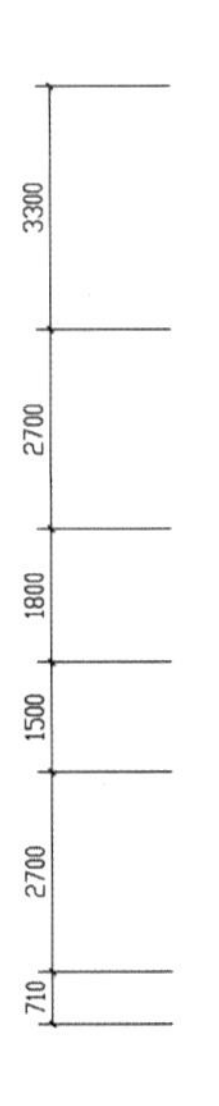

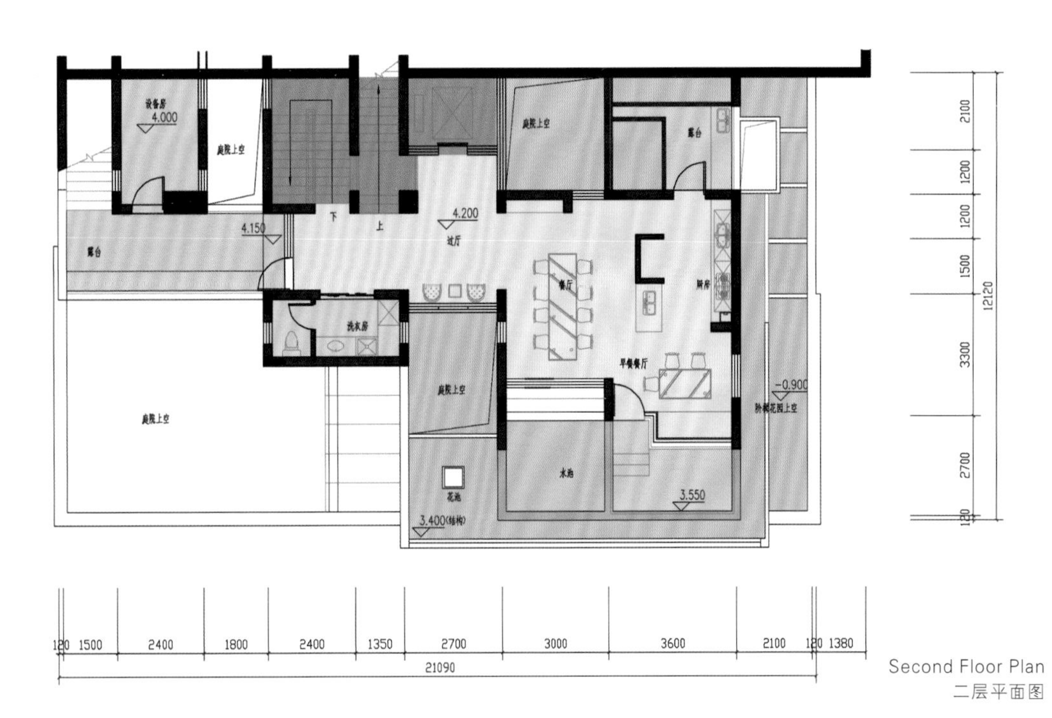

Second Floor Plan
二层平面图

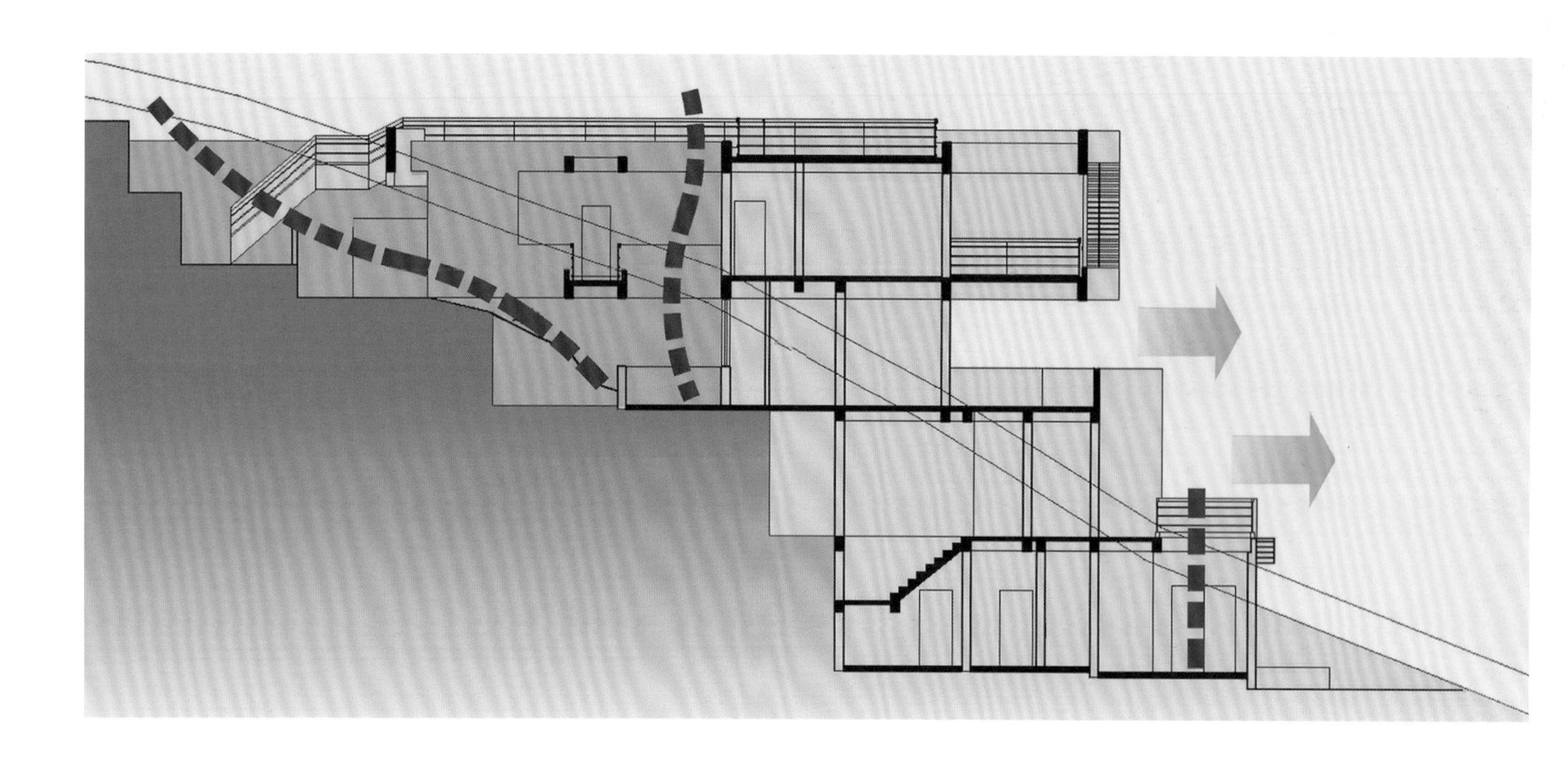

Heshan Ten Miles Radius

鹤山十里方圆

Location: Jiangmen, Guangdong, China
Developer: Fangyuan Group
Architectural Design: Huayang International Design Group
Gross Floor Area: 351,000 m^2
Plot Ratio: 0.4

项目地点：中国广东省江门市
开发商：方圆集团
设计单位：华阳国际设计集团
总建筑面积：351 000 m^2
容积率：0.4

FEATURES 项目亮点

The project uses the modern architectural language to abstract the essence of Chinese traditional architecture to create New Oriental Eclecticism Architecture and a Chinese poetic living space and let people re-experience the oriental restful and natural life described in the Chinese poetry.

项目用现代建筑语汇抽象中国传统建筑的精华，从而创造出一种新东方折衷主义建筑，营造中国山水诗画般的居住场景，让人重新体会到新东方人居中的"结庐在人境，而无车马喧"的生活境界。

KEYWORDS 关键词

New Oriental Eclecticism
新东方折衷主义

Chinese Courtyard
中式庭院

Artistic Conception of Landscape
山水意境

Overview

The project is in Heshan City of Guangdong Province and it backs the Dayan Mountains scenic spots. The project will be built in residential groups made up of independent, double, lined mansions and villas. It adopts the development idea of small towns and provides this new community with rich supporting resources.

项目概况

项目位于广东省鹤山市，背靠大雁山风景区。项目将建设成一个由独立大宅、双拼大宅、联排大宅和别墅等组成的住宅群，并采用小镇的开发理念，为新区提供丰富的配套资源。

Design Concept

In the design of Phase Two, Three and Four, the designer uses the modern architectural language to abstract the essence of Chinese traditional architecture, and proposes the idea of New Oriental Eclecticism Architecture, so that the future residents can fully experience the convenience of modern life, as well as learn the valuable heritage from the local culture. Extracting periodic features and local traditional living cultures and then focusing on the discussion of the meeting point of traditional and modern living cultures have provided a new clue for creating a modern oriental living style from planning to architecture space.

设计理念

在项目二、三、四期的设计中，设计师用现代建筑语汇抽象中国传统建筑的精华，提出新东方折衷主义建筑的理念，让将来的入住者既能充分体验到现代生活的便利性，又能从本地文化中吸取有价值的传承，提炼出既有时代特色又能体现本地传统的居住新文化，探讨传统居住文化与现代居住方式的契合点，从规划到建筑空间，都为创造现代东方人居模式提供新的线索。

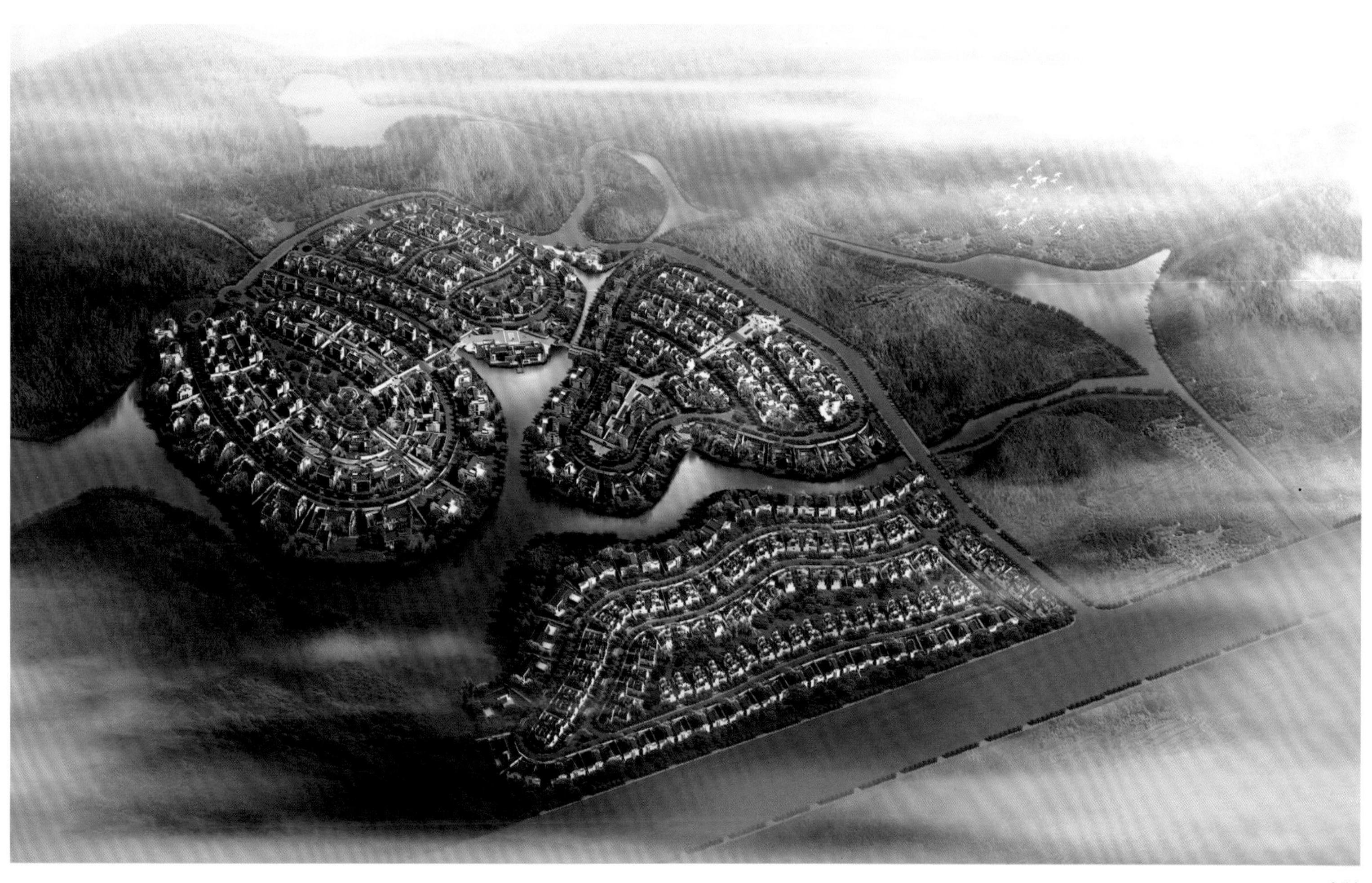

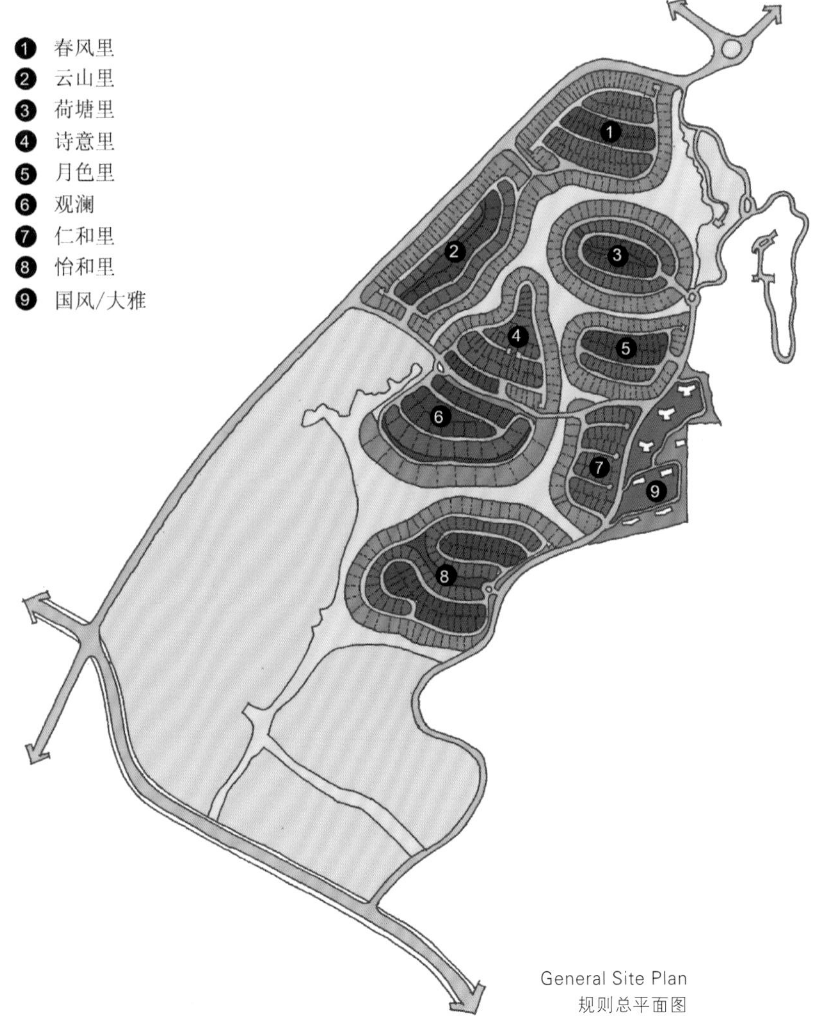

General Site Plan
规则总平面图

➤ Architectural Style

On the overall architectural style, the main color is based on white and warm gray, and the structure is mainly in modern design and embellished with the Chinese traditional element. It also uses the combination of materials such as traditional stone, brick, wood, spool tile and modern steel and glass to form a unique style.

➤ 建筑风格

在整体建筑风格的处理上，色彩以白色、暖灰色为主色调，造型以现代的设计手法组合形体的穿插为主线，提炼与点缀中国传统元素符号。利用传统材料如石材、砖、木材和圆筒瓦等和现代材料如钢材、玻璃等“混搭”，形成独具特色的造型风格。

➤ Architectural Details

On the architectural details, it extracts many of the Chinese classical architectural elements, and simplifies and represents them in modern ways. The facade learns from the characteristic of "Wharf Wall" of Huizhou residence and reduces overlapping in order to avoid visual complexity. The wall entrances use the common traditional entrance door and simplify it with modern steel and glass. The grid doors in the courtyard are evolved from the "Moon Gate" of Chinese classical garden, of which the circular openings and the real and false bricks not only separate the spaces, but also make the spaces penetrate each other and it's imaginative. The whole architecture creates a Chinese poetic living scene with "wall" "courtyard" "simplicity" "quietness" "village" and other elements, and lets people re-experience the oriental restful and natural life described in the Chinese poetry.

➤ 建筑细部

在建筑细部上，多处提炼中国古典建筑元素，并以现代建筑手法加以简化和表现。立面借鉴徽州民居中“马头墙”这一大特色，并结合体量简化跌落层次，避免过多跌级而造成视觉的繁复。院墙入口采用传统民居中常见的入户门楼形式，并运用现代材料钢和玻璃对其形式进行简化。庭院中的格构门从中国古典园林中的“月洞门”演化而来，圆形的门洞和虚实相间的砌块既分隔空间，又使得空间彼此渗透，引人遐想。整个建筑以“墙”“院”“素”“幽”“村”等元素营造出中国山水诗画般的居住场景，让人重新体会到新东方人居中的“结庐在人境，而无车马喧”的生活境界。

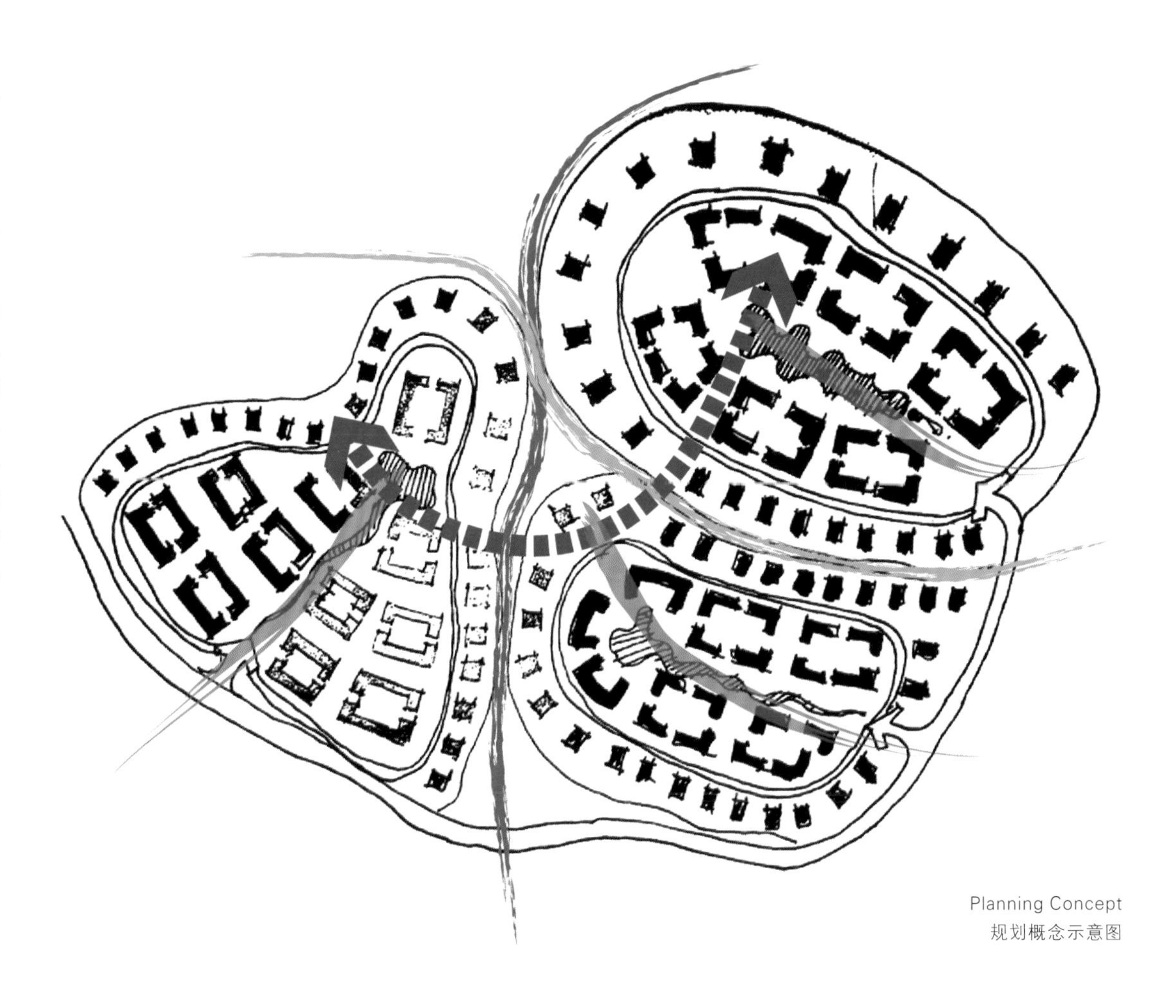

Planning Concept
规划概念示意图

➤ House Type Design

The house type design, according to the specific community characteristics of Chinese architecture, emphasizes the transition of space and sense of layering and sequence. In the meantime, the use of Chinese courtyard strengthens the Chinese flavor, and the combination and separation of the walls enrich the spatial levels.

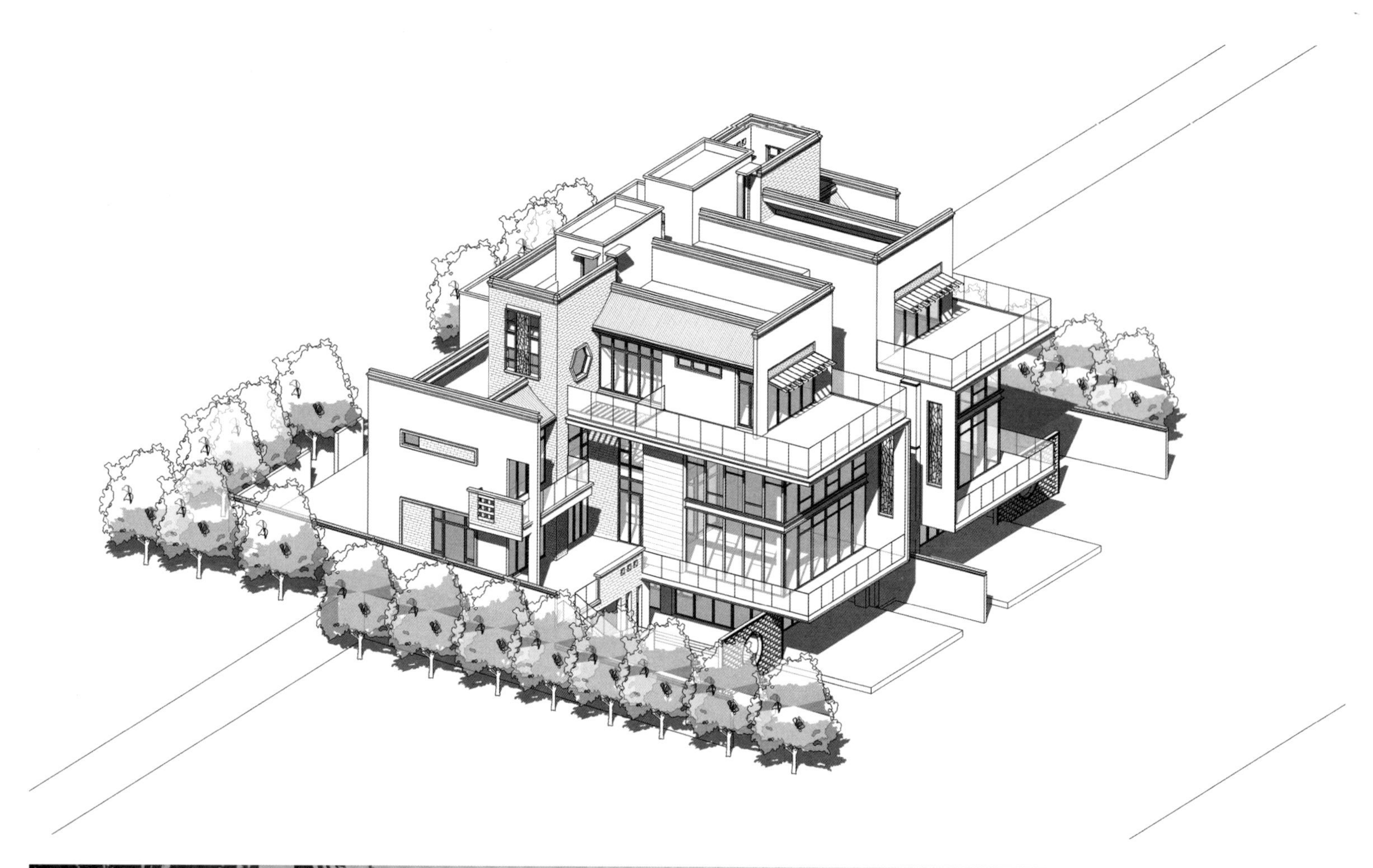

➤ 户型设计

在户型设计上，根据中国建筑特有的社区特点，强调空间的过渡、转折以及层次感和序列感。同时，中式庭院的使用强化了中国韵味，其中院墙的灵活组合与分隔又使得空间层次更加丰富。

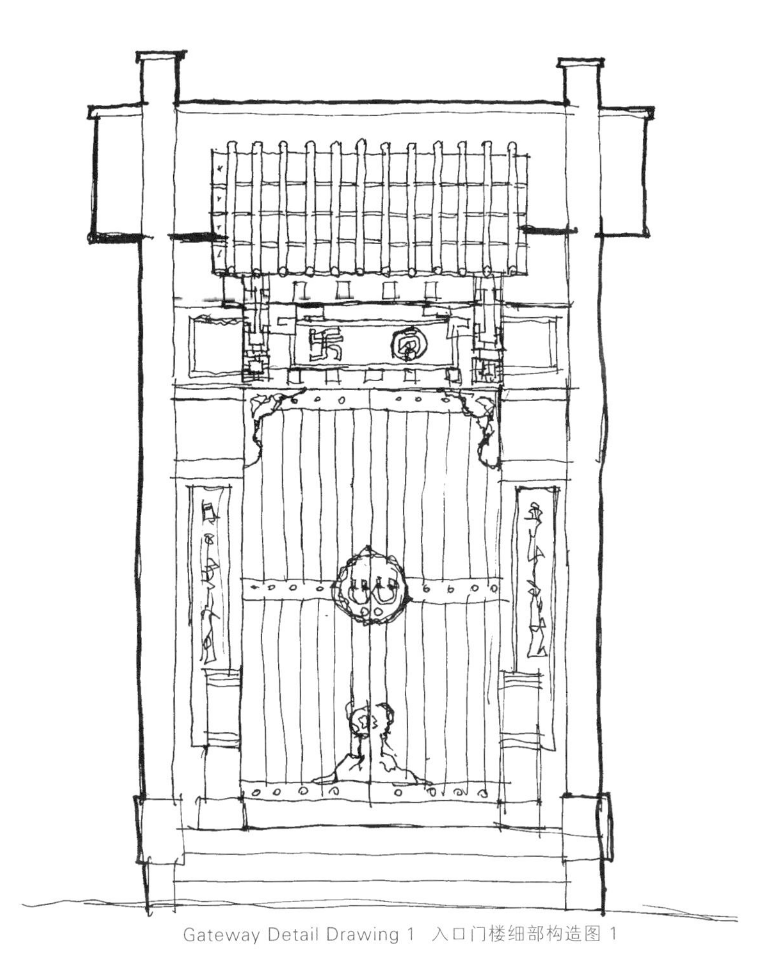

Gateway Detail Drawing 1　入口门楼细部构造图 1

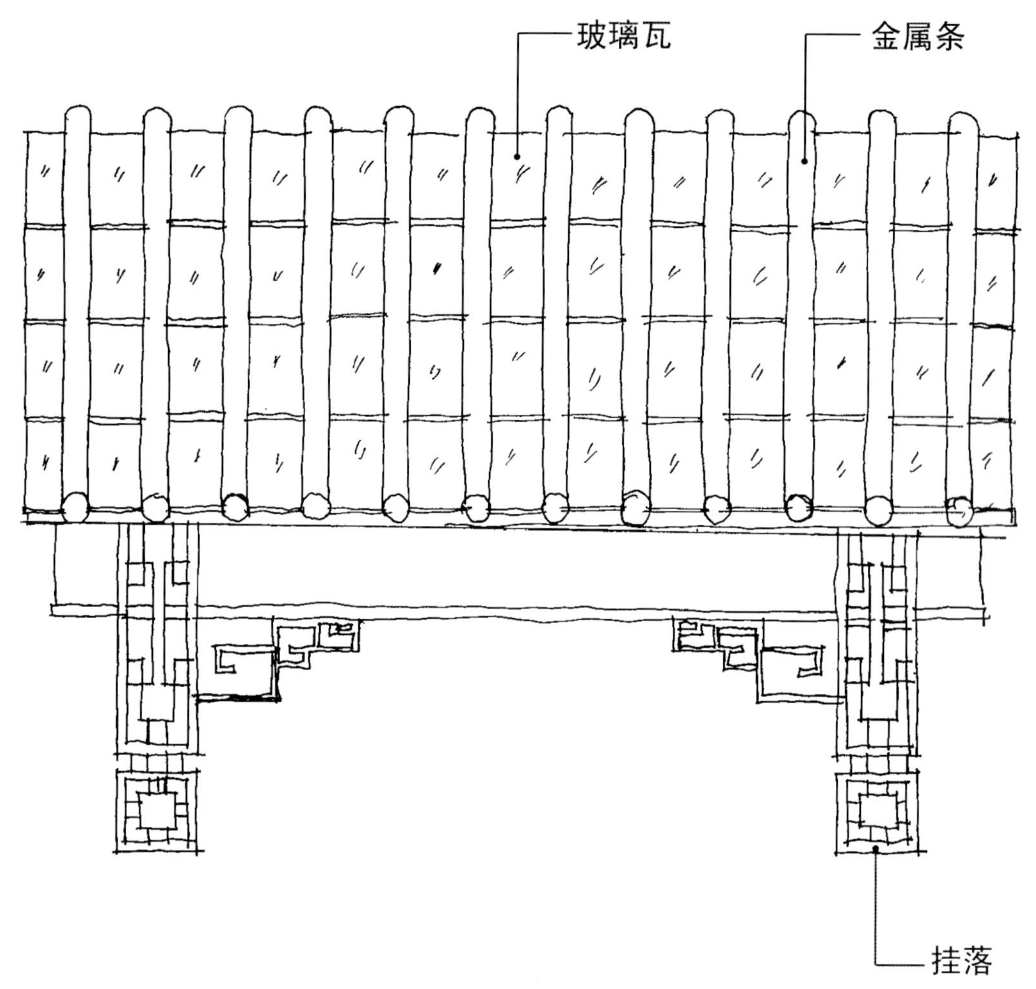

Gateway Detail Treatment
入口门楼细部构造处理

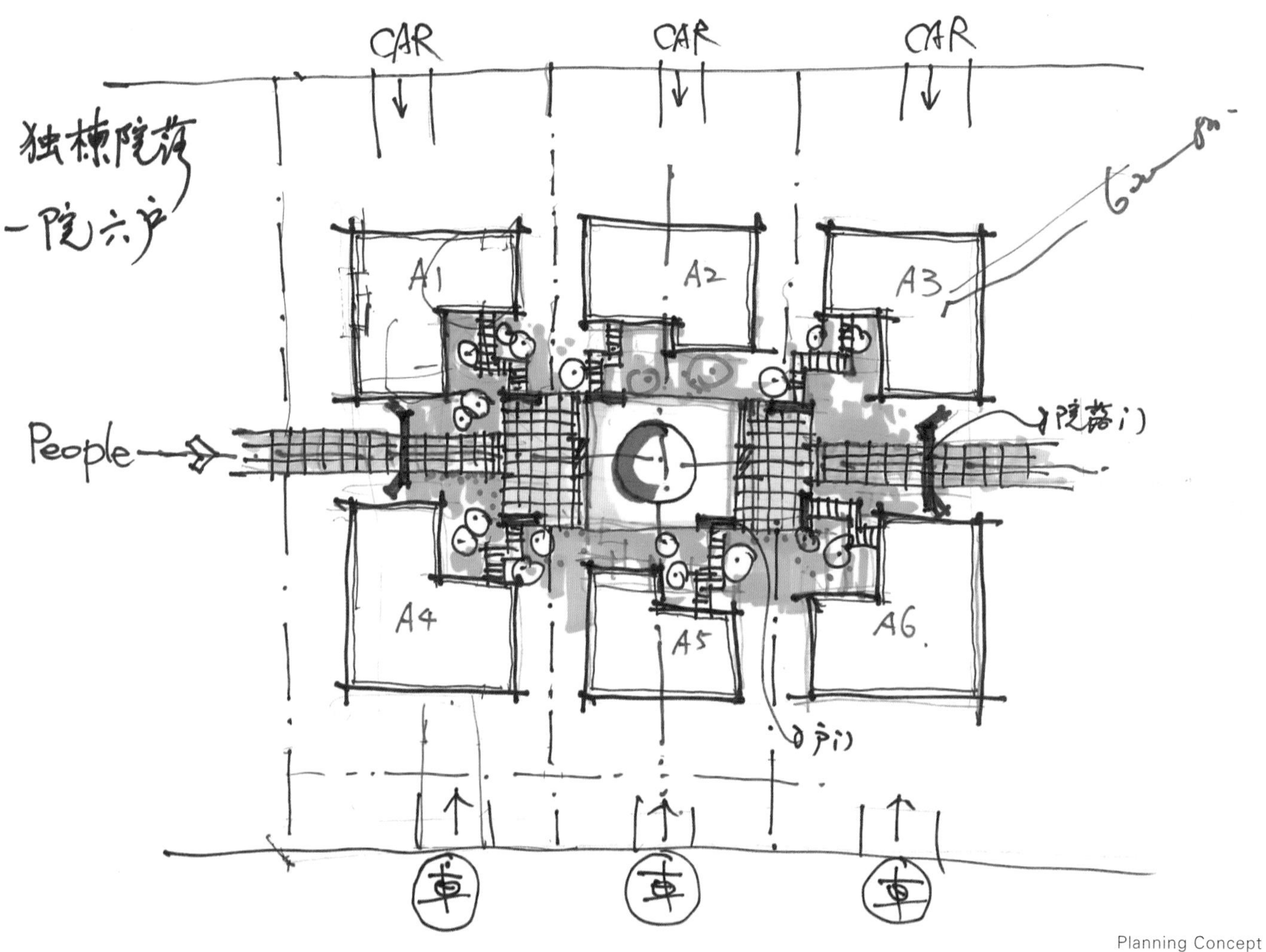

Planning Concept
规划概念示意图

上檀居

上檀居

Xi'an OCT • 108 Block
西安华侨城 · 壹零捌坊

Location: Xi'an, Shaanxi, China
Developer: Xi'an OCT Properties Co., Ltd.
Architectural Design: Shenzhen Huahui Design Co., Ltd.
Gross Floor Area: 174,400 m^2
Plot Ratio: 1.0
Awards: Prize for Chinese Residential Area Architectural Design in 2012
The Comprehensive Awards of National Classic Architectural Planning and Design Contest in 2013

项目地点：中国陕西省西安市
开发商：西安华侨城房地产开发有限公司
设计单位：深圳华汇设计有限公司
总建筑面积：174 400 m^2
容积率：1.0
获奖情况：2012年华人住宅与住区建筑设计奖
2013年全国经典建筑规划设计方案竞赛综合大奖

KEYWORDS 关键词

Chinese Architecture
中式建筑

Lifang-style Planning
里坊式规划

Liuhe Courtyard Villas
六合院别墅

FEATURES 项目亮点

On the facade design, it shows a new Chinese style by extracting the classical Chinese architectural language and using modern constructional materials and structures.

立面设计上，通过提炼中式建筑的典型形式语言，运用现代的建筑材料与构造方式，演绎了一种新中式格调的建筑风格。

Overview

Xi'an OCT • 108 Block, with its location opposite to the east of Datang Furong Garden and on the north of Tang City Wall Relic Park, is in the core area of Qujiang New District, Xi'an City. It is a low density community with complex terrains which are mostly continuous slopes that rise step by step from northwest to southeast.

项目概况

项目地处西安市曲江新区核心区，大唐芙蓉园东侧对面，唐城墙遗址公园北面。项目属于低密度社区，地形较复杂，呈台地地貌，大部分为连续坡面，从西北到东南逐级抬高。

Design Concept

1. It learns and borrows from Chinese traditional urban Lifang planning style, and emphasizes the rank and order of space to create a progressive space of multiple levels.

2. On the formation mode of the architecture, it uses compound courtyard as the main construction unit and organizes the courtyard as the core of space to represent the advocation of inner space by the Chinese or even the orientals.

3. It uses the gentle slope and builds the terrace appropriately to make the architecture and the site mix together.

4. The overall layout fully uses and discovers the advantages of the natural resources and the terrain features of the base, and optimizes the relations of the architecture and the outer space, so that the architecture and the environment can be a harmonious unity.

5. It deduces the artistic conception of oriental architecture through modern material language.

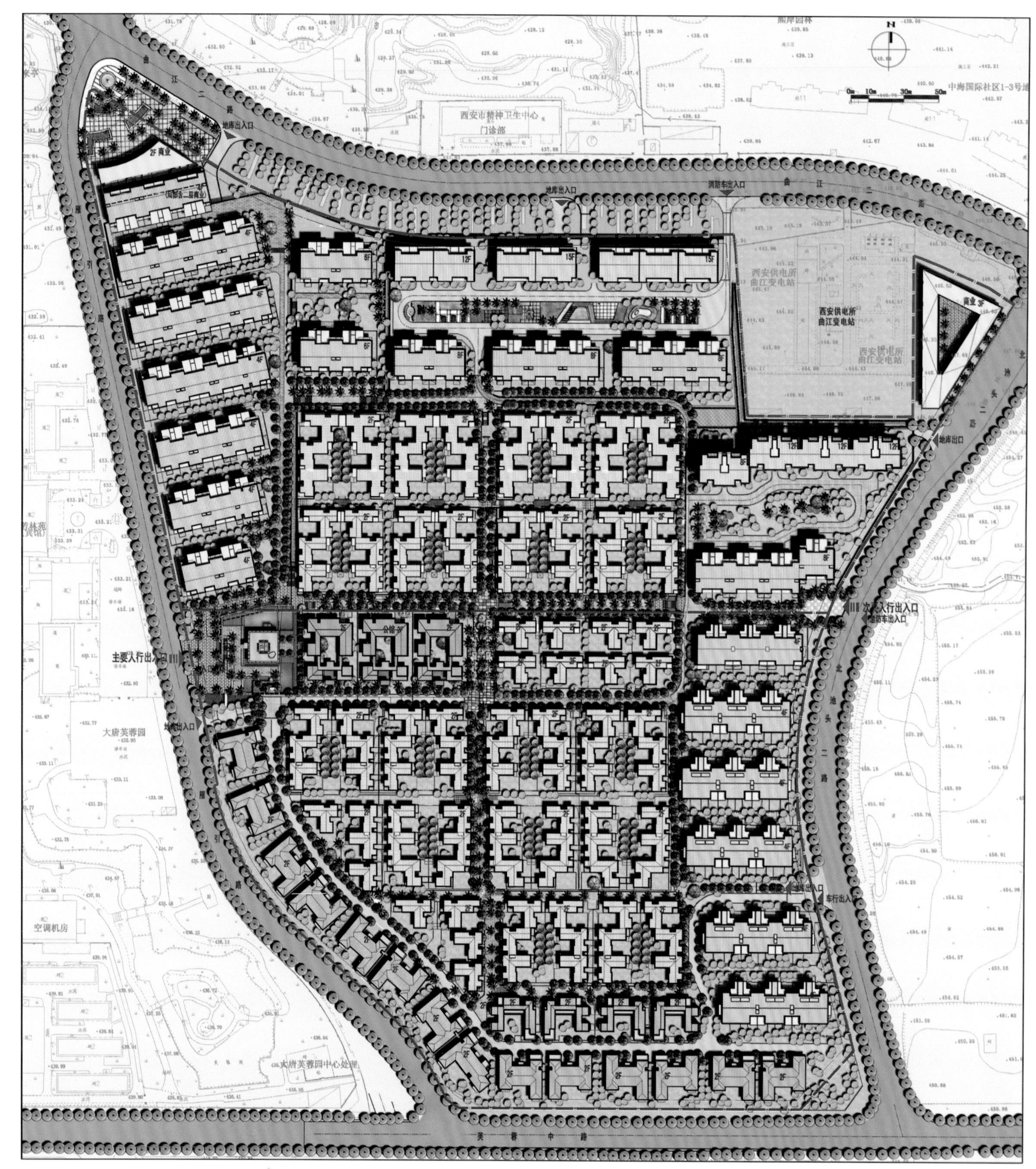

Site Plan
总平面图

➤ 设计理念

1.吸取并借鉴了中国传统城市的里坊式规划模式，强调空间的等级与秩序，营造出多层次的递进空间。

2.在建筑组成模式上，以合院为主要的建筑单元，以院落为主要的空间核心进行组织，表现中国乃至东方人对这种内向空间的崇尚。

3.利用缓坡地形，合理地营造台地，将建筑与场地融合。

4.整体布局充分利用和发掘基地周边的自然资源优势与基地地形特点，优化建筑与外部空间的关系，使建筑与环境形成和谐统一的整体。

5.通过现代材料语言，演绎东方建筑意境。

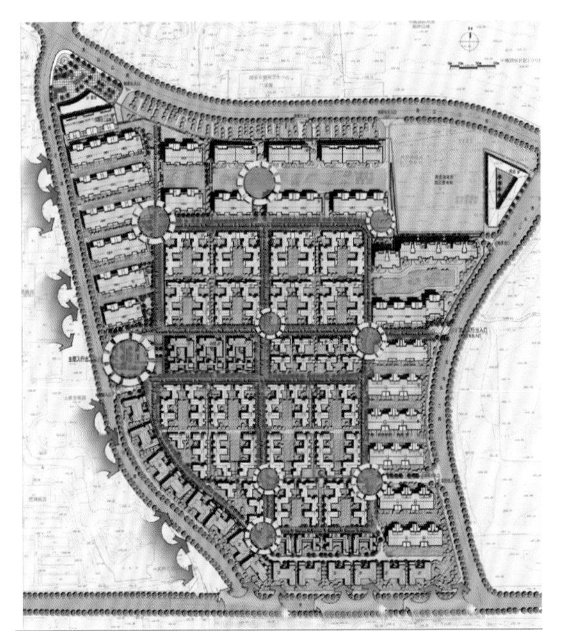

Analysis Drawing
景观分析图

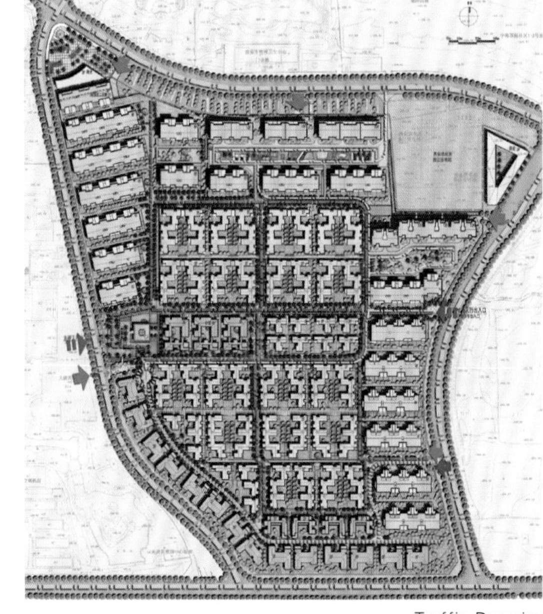

Traffic Drawing
交通分析图

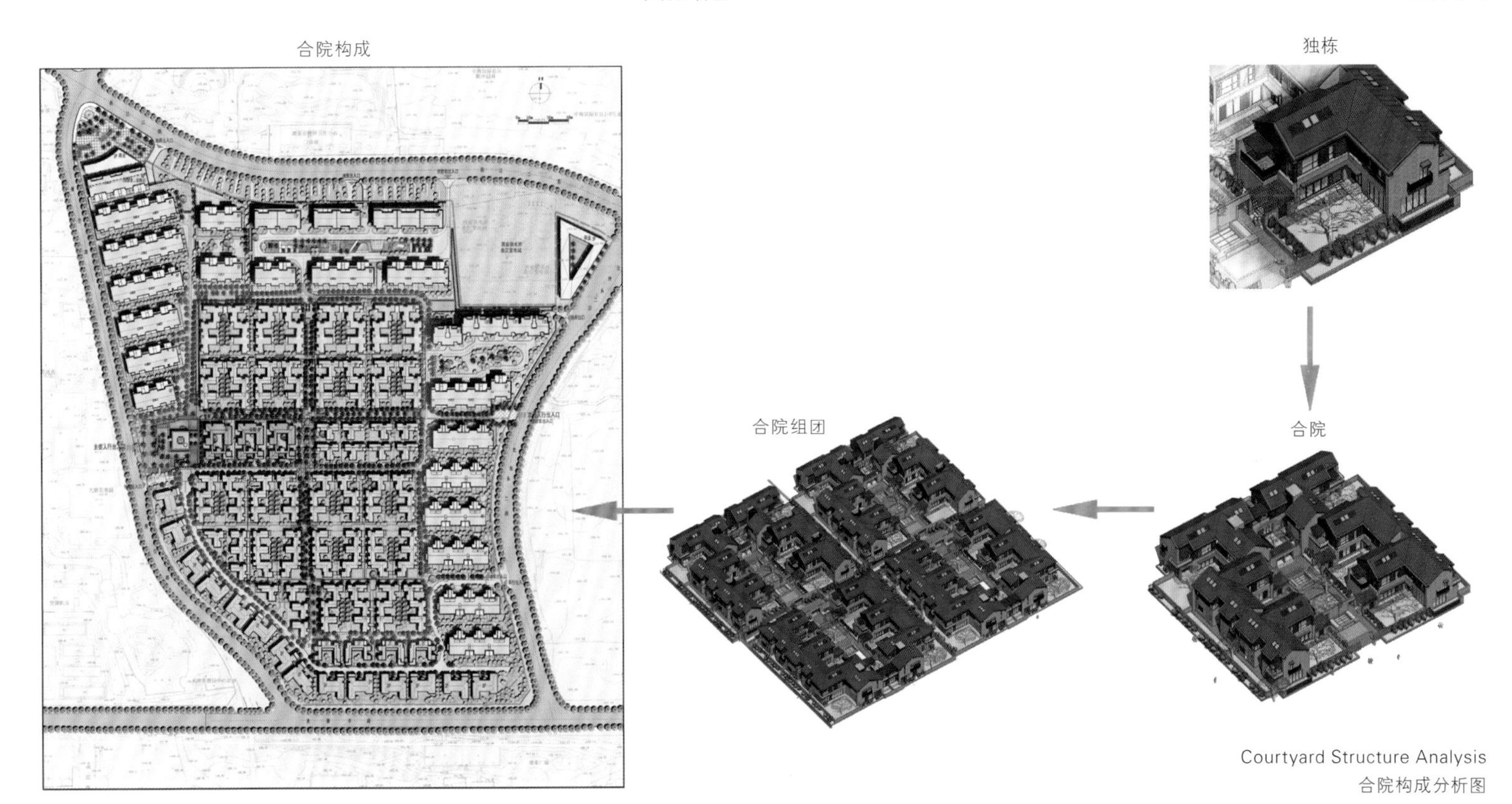

Courtyard Structure Analysis
合院构成分析图

Planning Layout

The central west of the land is the horizontal axis starting point of the community landscape. Walking through the square, club and then the terrace garden, you can get into the whole community.

The whole planning structure respects the local tradition and residential combination mode. It emphasizes the rank and order of space with the north-south cross grid as the main space structure. It displays the traditional Lifang planning from street to alley, public courtyard, private courtyard and finally to the home.

Meanwhile, in order to restore the quiet cultural environment, the main traffic for vehicles is led into underground, while on the ground is mainly the pedestrian traffic. Every house can be accessed easily through the underground traffic, and it combines the traditional living atmospheres and the request of modern people for fast and convenient life organically.

规划布局

用地西侧中央为社区的景观横轴起点，通过广场—会所—台地花园，进入整个社区。

整体规划结构尊重本地传统及民居组合模式，以南北十字网格布局为主要空间构架，强调空间的等级与秩序，街—巷—公共院落—私家院落—家，层层推进，将传统里坊式规划浓缩体现在这里。

同时，为了还原出静谧的人文环境，将社区的主要车行交通引入地下，地面上主要是人行交通。地下交通可以很便利地接引每户，将传统的生活情调与现代人讲求便利快捷的生活要求有机地结合。

South Elevation
南立面图

North Elevation
北立面图

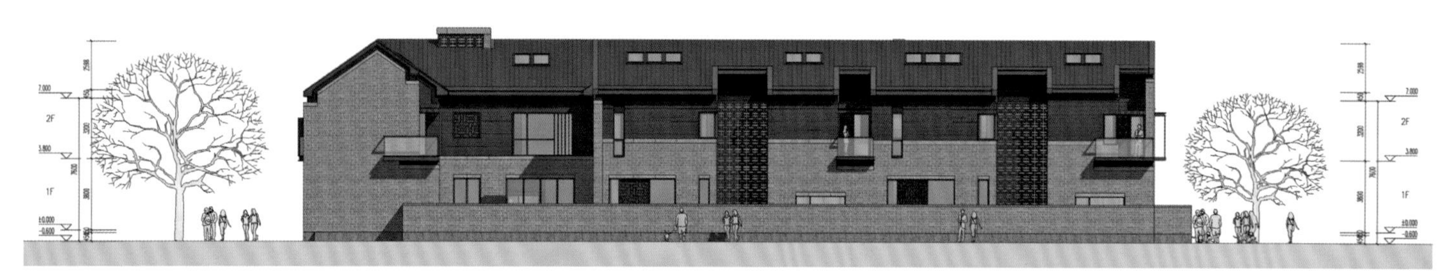

West Elevation
西立面图

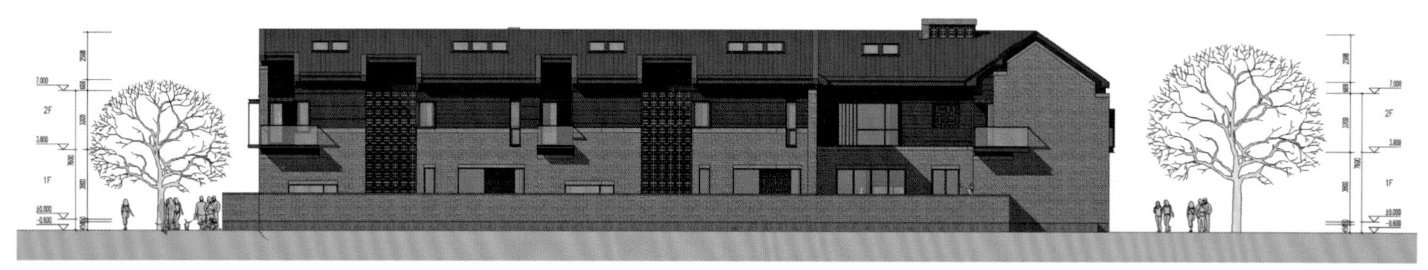

East Elevation
东立面图

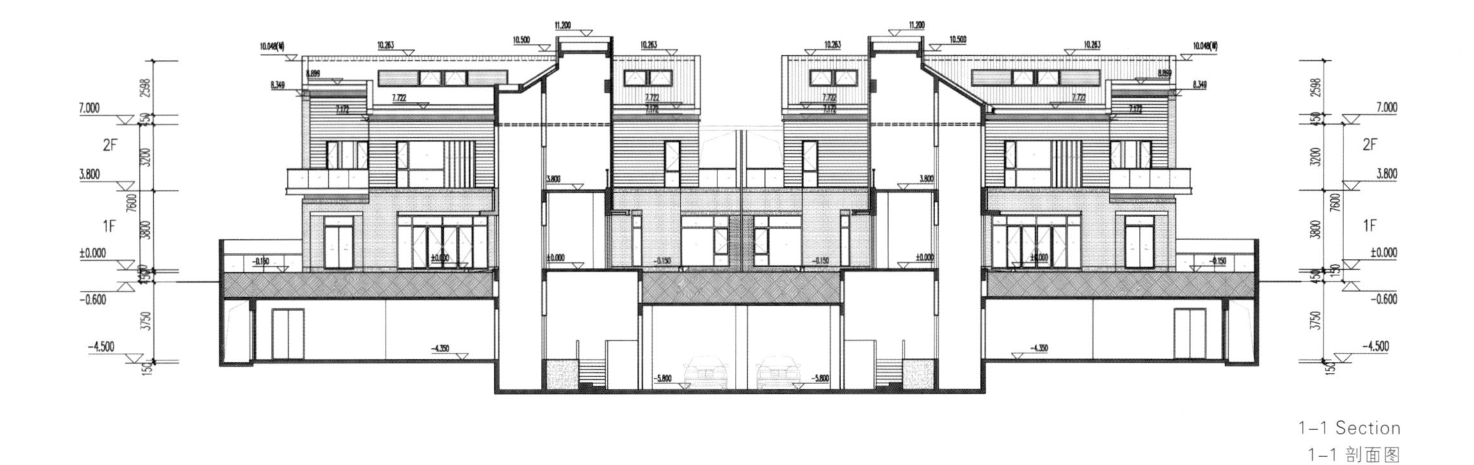

1–1 Section
1–1 剖面图

2–2 Section
2–2 剖面图

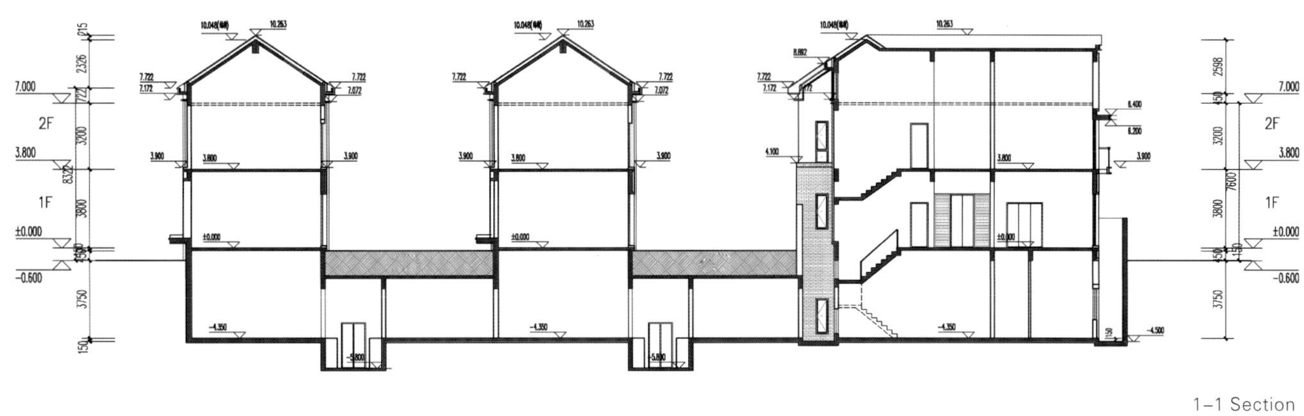

1–1 Section
1–1 剖面图

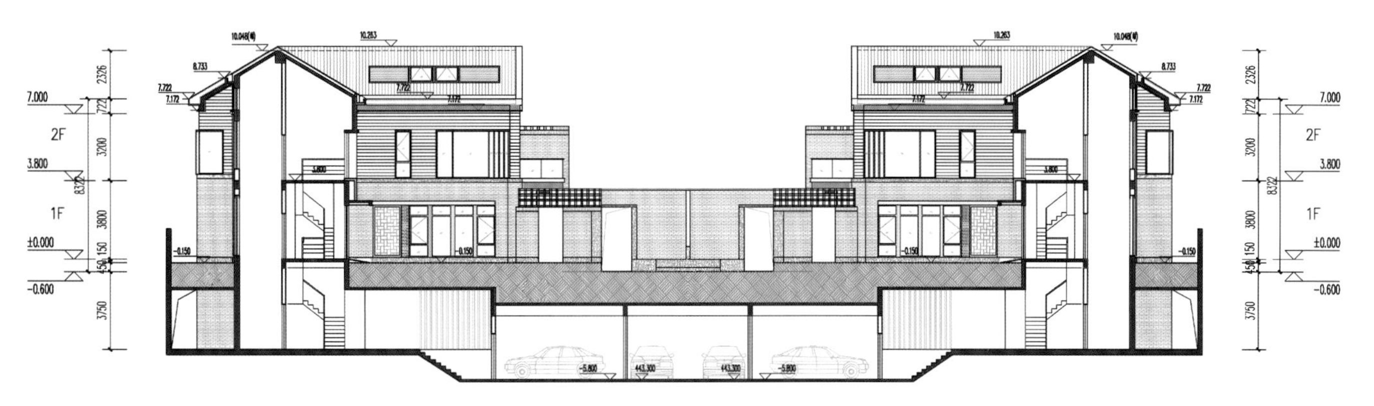

2–2 Section
2–2 剖面图

➤ Architectural Design

The designer focuses on the shaping of new compound courtyard mode, which is made up of six combined courtyard villas. The spatial system of each courtyard is level by level, from two public courtyards, then to the private courtyard and finally to the home. The integration of inward private courtyard and semi-public courtyard gives a new life to the courtyard. The private courtyard is the center of every family life and the extension of traditional residential space. Public courtyard is the shared communication space of the six residences. It is also the unit of the community life and has been a part of the modern urban life.

In consideration of the characteristics of the northern climate, the unit building has fully satisfied the request for lighting and ventilation by the residents. It has abandoned the physical loss in function of the traditional courtyard, and emphasized the combination of traditional space and modern life. It focuses on the control of spatial scale and controls the building height within 2 floors so as to create a pleasant courtyard scale.

On the facade design, it shows a new Chinese style by extracting the classical Chinese architectural language and using modern constructional materials and structures.

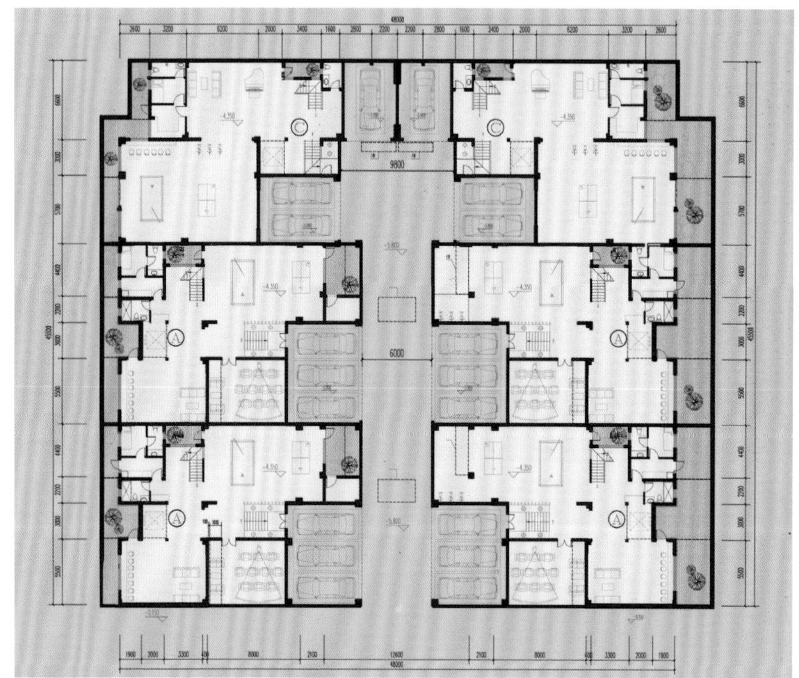

Basement Floor Plan 地下层平面图

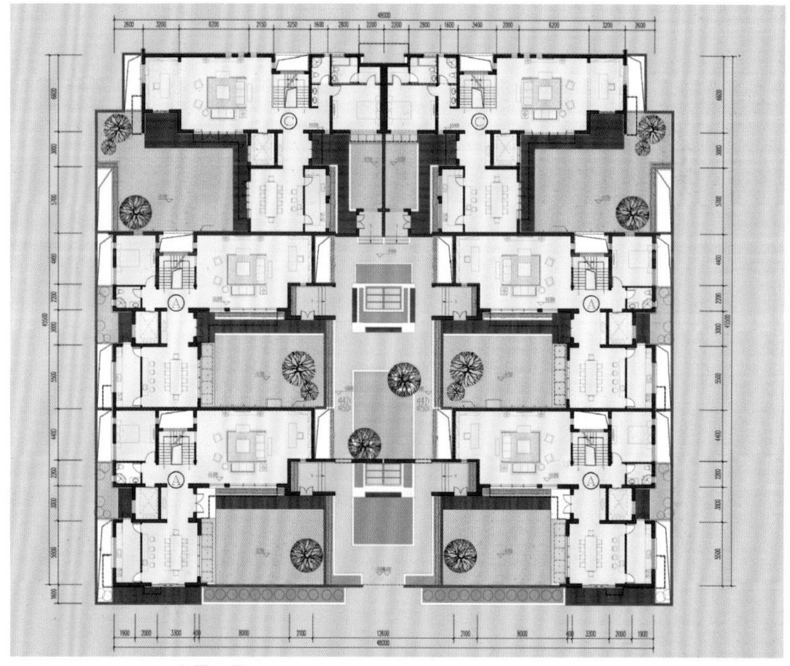

First Floor Plan 一层平面图

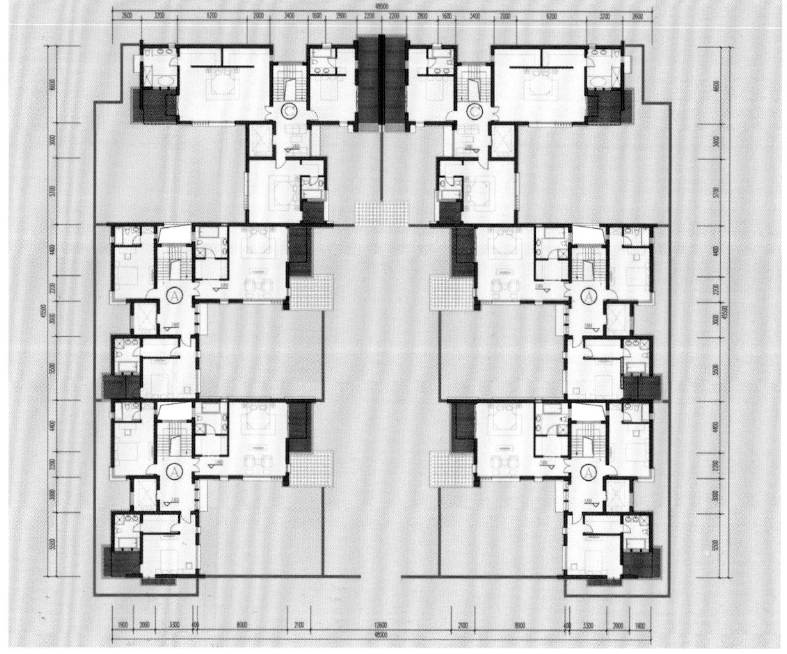

Second Floor Plan 二层平面图

➤ 建筑设计

在本项目中，设计师重点塑造了新型的合院模式——由六户有机组成的“六合院”别墅。每个合院都是一个层进式的空间体系：两进公共庭院—私家院落—家。内向性私家庭院与半公共庭院的融合赋予了院落新的生命。私家庭院是每户家庭生活的中心，是传统住宅空间的延续；公共庭院是六户共享的交流场所、社区生活的单元，是现代城市生活的一部分。

考虑到北方的气候特征，建筑单体充分满足了当地居民对采光、通风等方面的需求，摒弃了传统合院在居住功能上的物理性缺失，强调传统空间与现代生活的结合。注重空间尺度的控制，将建筑高度控制在两层以内，以塑造宜人的庭院尺度。

立面设计上，通过提炼中式建筑的典型形式语言，运用现代的建筑材料与构造方式，演绎了一种新中式格调的建筑风格。

Kaifeng Water System, Xihe Mansion, Henan

河南开封水系熙和府

Location: Kaifeng, Henan, China
Architectural Design: Shenzhen General Institute of Architectural Design and Research Co., Ltd.
Land Area: 14,910 m^2
Gross Floor Area: 20,605 m^2
Plot Ratio: 1.36
Building Density: 34%
Green Coverage Ratio: 30%

项目地点：中国河南省开封市
设计单位：深圳市建筑设计研究院总院有限公司
占地面积：14 910 m^2
总建筑面积：20 605 m^2
容积率：1.36
建筑密度：34%
绿化覆盖率：30%

KEYWORDS 关键词

Slope Roofs of Song Dynasty Style
宋式坡屋顶

Visual Corridor
视觉通廊

Greening System
绿化系统

FEATURES 项目亮点

The overall architectural style and features are in the form of slope roofs of Song Dynasty style, responding to water system phase II landscape engineering. The design emphasizes proportions, lines and scales, pursuing perfect building facade and shape.

整体建筑风貌采用宋式坡屋顶的形式，与水系二期景观工程相呼应。建筑外观设计注重比例、线条及尺度，追求建筑立面及形体的完美。

➤ Overview

The water system phase II engineering is urban construction, located in old Kaifeng. It boasts advantageous geographic location, the subways spreading along the two sides of phase II water system. It is a key project for Kaifeng, Henan. The project is a complex building group, including residence, travel, leisure, shopping, catering, culture, entertainment, communication, business and other functions. Xihe Mansion is located in the south of phase II water system engineering, with convenient transportation and good environment.

➤ 项目概况

水系二期工程为旧城改造，是河南省开封市的重点项目，位于开封市老城区。地理位置优越，所开发的地块分布在二期水系河道两侧。项目是集合居住、旅游、休闲、购物、餐饮、文化、娱乐、社交和商务等功能于一体的综合建筑群。熙和府项目位于水系二期工程南侧，交通便利，环境优美。

➤ Overall Planning

Buildings of the community are designed in the form of multi-layer strip and set in several rows, highlighting the neat and concise design concept.

➤ 总体布局

小区内的建筑形式为多层条式，采用行列式的建筑布局，突出整齐、简洁的设计理念。

➤ Architectural Design

The overall architectural style and features are in the form of slope roofs of Song Dynasty style, responding to water system phase II landscape engineering. The design emphasizes proportions, lines and scales, pursuing perfect building facade and shape.

As for the use of color, following the principle of high brightness and low chroma and combining with land planning around Kaifeng water system, the designers have built warm and beautiful houses with lively overall shape and created a good residential environment.

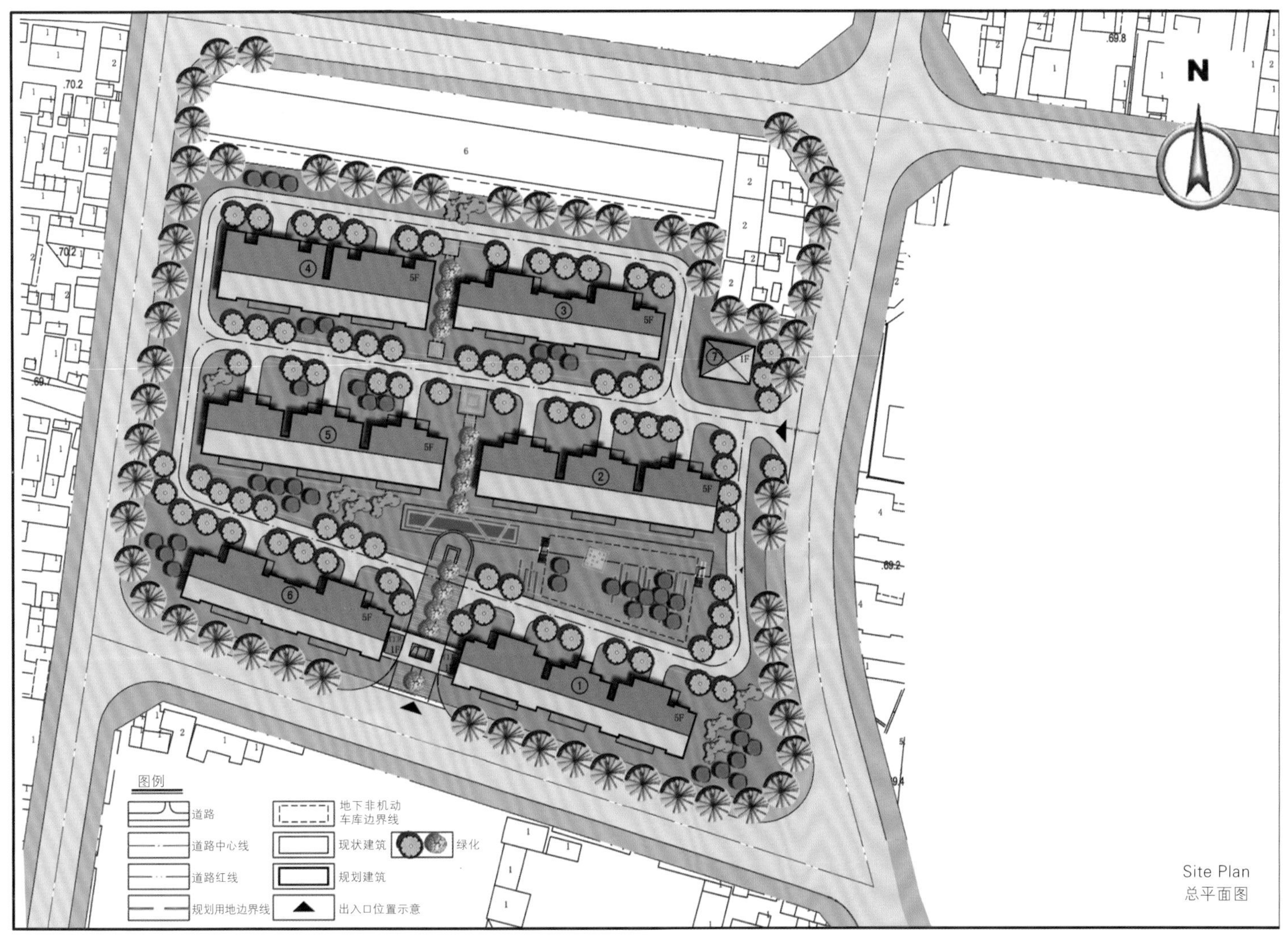

Site Plan
总平面图

➤ 建筑设计

整体建筑风貌采用宋式坡屋顶的形式，与水系二期景观工程呼应。建筑外观设计注重比例、线条及尺度，追求建筑立面及形体的完美。

在色彩上，遵循高亮度低彩度原则，并结合开封水系周边的规划，建造整体造型明快、温暖靓丽的住宅，创造良好的居住环境。

➤ Floor Plan Design

The project brings in advanced home design concept and adds in new materials, new process and new technology to create excellent architectural physical environment and living space.

➤ 户型平面设计

项目引进超前的居家设计理念，加入新材料、新工艺、新技术元素，力求创造出优秀的建筑物理环境和生活空间。

➤ Landscape Design

Entrance zone: complete the axis landscape design at the entrance to form an integrated and smooth visual corridor. Entrance design should have ornamental value and identification in space image, becoming the window that displays the whole community's quality and image.

Central active zone: cross intersection area of two landscape green axes, landscape pavilion, landscape wall, small water surface, and hard flooring are set there to build a central landscape plaza as the visual focus.

Green space system: the residential green space is divided into three parts, including public green space, green space between houses and green belts along the roads. Taking the public green space as the core and connecting each green space between houses through the green belts along the roads to create multi-form and multi-level spacial greenbelt system.

➤ 景观设计

入口区：做好入口处的轴线景观设计，形成完整流畅的视觉通廊。入口的设计在空间意象上应具有观赏性与标识性，成为展示小区品质与形象的窗口。

中心活跃区：在两道景观绿轴的十字相交区域，布设景观亭、景观墙、小水面和硬地板，打造中心景观广场，使其成为景观设计的视觉焦点。

绿地系统：小区绿地分为公共绿地、宅间绿地和沿路绿化带三个部分。以公共绿地为核心，通过沿路绿化带串联起各个宅间绿地，形成多形态、多层次的空间绿地系统。

South Elevation of Building 2 & 5
2、5 栋南立面图

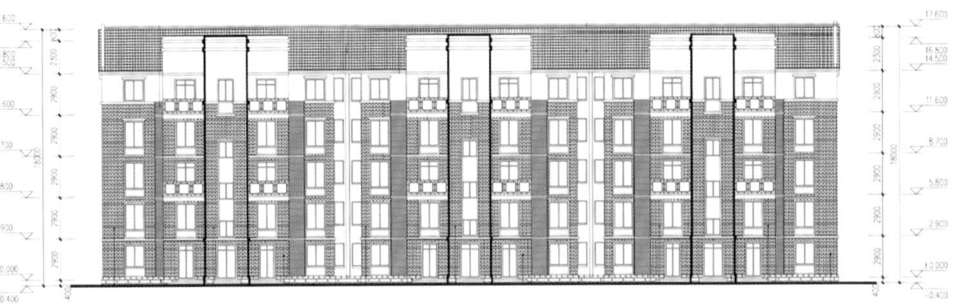

North Elevation of Building 2 & 5
2、5 栋北立面图

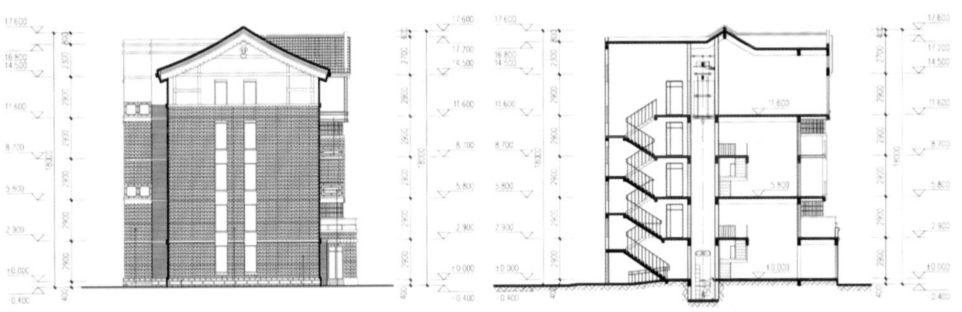

West Elevation of Building 2 & 5
2、5 栋西立面图

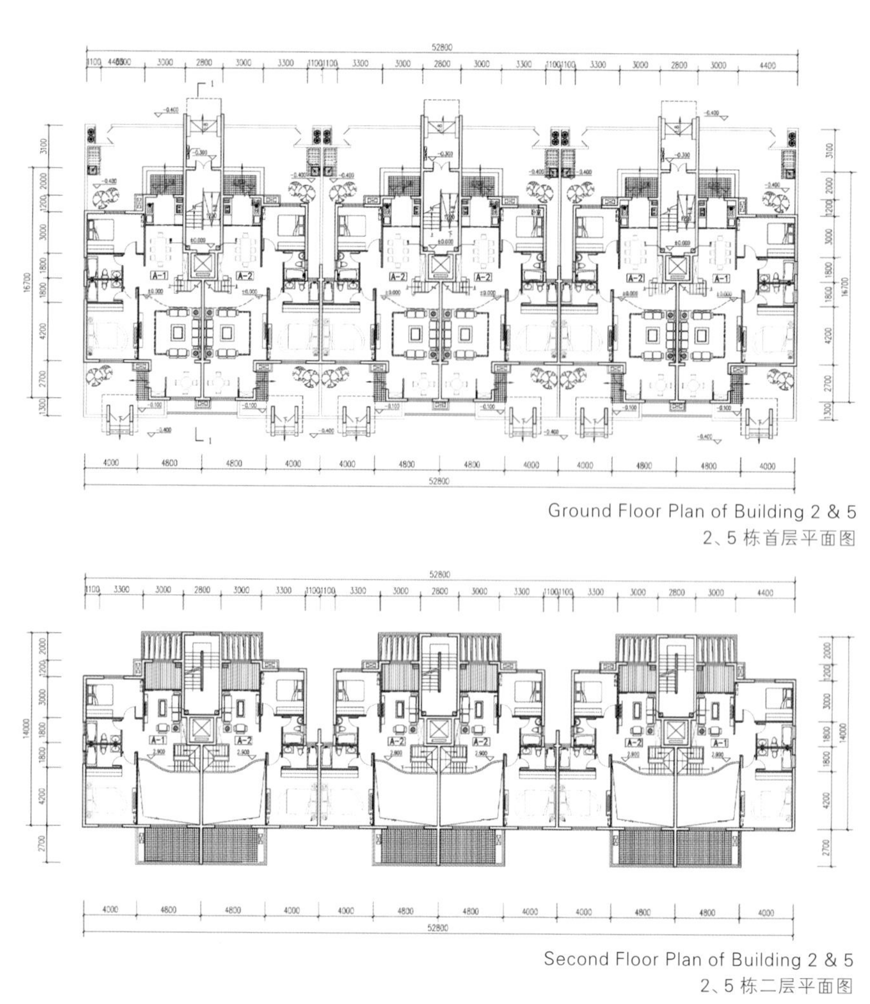

Ground Floor Plan of Building 2 & 5
2、5 栋首层平面图

Second Floor Plan of Building 2 & 5
2、5 栋二层平面图

Liuzhou Shengtian Longwan

柳州盛天龙湾

Location: Liuzhou, Guangxi Zhuang Autonomous Region, China
Architectural Design: Shing & Partners
Cooperator: Liuzhou Shengtian Real Estate Development Co., Ltd.
Land Area: 120,000 m^2
Floor Area: 235,000 m^2
Awards: 1st Prize of Guangzhou Excellent Design in 2012

项目地点：中国广西壮族自治区柳州市
设计单位：汉森伯盛国际设计集团
合作单位：柳州盛天房地产开发有限公司
占地面积：120 000 m^2
建筑面积：235 000 m^2
获奖情况：2012年广州市优秀设计一等奖

FEATURES 项目亮点

Architectural style of the project is in modern and simple Chinese style, so it features Chinese architectural elegant style and modern and simple artistic expression style.

项目建筑风格为现代简约中式，既具有中式建筑的典雅风格，又具备现代简约的艺术表现风格。

KEYWORDS 关键词

Simple Chinese Style
简约中式

Elegant Style
典雅风格

Water System Landscape
水系景观

➤ Overview

The project is located at the junction of Yanghe Avenue and Jinglan Bridge in Yanghe District, Liuzhou City, surrounded by mountains with beautiful natural environment, west to Liujiang, east to the Yanghe Overpass, close to the exit of Guilin-Liuzhou Expressway, lying by the vital transportation line of Liuzhou, Guilin and Nanning, and only a 10-minute drive to the city center. The project has a gross floor area of about 240,000 m^2, and a greening landscape area of about 80,000 m^2, and the building density is 12% (including green belt). There are 1,800 sets of residence and more than 900 underground parking spaces in the community, which is composed of 20 blocks of high-rise buildings with 15-28 layers, of which two buildings of 28 layers, six buildings of 25 layers, one building of 20 layers, ten buildings of 18 layers and one building of 15 layers. One of the 25-layer buildings is an independent apartment. The project is also equipped with a kindergarten and community business.

➤ 项目概况

项目位于柳州市阳和区阳和大道与静兰大桥交界处，周边群山环抱，自然环境优美，西眺柳江，东临阳和立交桥，紧临桂柳高速出口，处于柳州进出桂林、南宁的交通要道，距市中心仅10分钟车程。项目总建筑面积约240 000 m^2，绿化景观用地面积约80 000 m^2，建筑密度12%（含绿化带），小区规划有1 800套住宅，地下车位900多个。由20栋15~28层高层建筑组成，其中两栋28层、六栋25层、一栋20层、十栋18层、一栋15层，其中一栋为独立管理的25层公寓式住宅，配套有幼儿园、小区商业。

➤ Architectural Design

Architectural style of the project is in modern and simple Chinese style, so it features Chinese architectural elegant style and modern and simple artistic expression style. The whole project is composed of high-rise buildings. House type includes one bedroom, two and three bedrooms, and double deck, with the area between 64 m^2 and 219 m^2. The ground of lobby in the first floor is paved by mosaics granite, and the wall by marble. The lobby is hung with elegant and noble ceiling and lamps.

Site Plan
总平面图

➤ 建筑设计

项目建筑风格为现代简约中式，既具有中式建筑的典雅风格，又具备现代简约的艺术表现风格。整个项目由高层组成。户型设计有一房、两房、三房、楼中楼，面积为64~219 m^2。首层入口大堂地面铺马赛克花岗石，墙身则为大理石。大堂悬挂着典雅、尊贵的天花板和灯饰。

➤ Landscape Design

The project has a public green belt of 40,000 m^2 and 5 groups of gardens and water system landscapes. The whole project's central axis is magically connected by various water system landscapes, which blends all the community's landscapes into a harmonious whole. In the area of community along the river, it is planned to build a municipal riverside park, which holds the best river view and riverside ecological leisure area, enjoying more obvious advantage of landscape.

➤ 景观设计

项目小区拥有40 000 m^2的公共绿化带，5个组团园林水系景观。整个项目中轴由各个水系景观相连接，水系园林虚实相连，使得整个小区园林景观浑然一体。小区临江面将规划一个市政府江滨公园，拥揽一线江景及滨江生态休闲区，景观优势更加明显。

小俩口
青苹果
晒幸福
帮了我！
2600000

创国家环保模范

创国家环保模范城

Luoyang Century City

洛阳世纪华阳

Location: Luoyang, Henan, China
Owner: Luoyang China Asia Properties Limited
Architectural Design: Shenzhen General Institute of Architectural Design and Research Co., Ltd.
Design Team: Yang Xu, Peng Ying, Zhang Xusong, Feng Zhiyong, Deng Xiaoshun, Xia Guang, Wang Xiaoliang
Gross Land Area: 290,171 m^2
Gross Floor Area: 1,054,107.97 m^2
Plot Ratio: 2.84
Green Coverage Ratio: 37%

项目地点：中国河南省洛阳市
项目业主：洛阳中亚置业有限公司
设计单位：深圳市建筑设计研究总院有限公司建筑创作院
设计人员：杨旭　彭鹰　张绪松　冯志勇　邓晓舜　夏光　王晓亮
总占地面积：290 171 m^2
总建筑面积：1 054 107.97 m^2
容积率：2.84
绿化覆盖率：37%

KEYWORDS 关键词

Traditional Architecture
传统建筑

Chinese Feature
中式特色

Openess
开放性

FEATURES 项目亮点

The project design of architectural modelling appropriately uses greyish white color tone of traditional Chinese architectural form, and creates a modern community architectual image with Chinese feature through application of modern architectural materials.

项目在建筑造型的设计上适当运用了中国传统建筑形式的灰白色调，利用现代的建筑材料营造出具有中式特色的现代社区建筑形象。

Overview

The project is located in the central part of Jianxi District, Luoyang City, with a gross land area of 290,171 m^2. Yan'an Road on the north and Jiudu Road on the south both are the main traffic arteries in the city, and the site is adjacent to the West Park on the southwest. Land for the project is divided into block A, B and C three parts by the city roads. Block A and C each occupies a piece of urban public green space. The entire building group consists of a commercial office building, several commercial apartments, three sets of residential area and a set of business building. The gross floor area reaches to 1,054,108 m^2.

项目概况

项目位于洛阳市涧西区中心位置，总占地面积290 171 m^2。基地北面的延安路与南面的九都路均为城市主要交通干道，且西南临西苑公园。项目用地被城市道路划分为A、B、C三个地块，其中A、C地块各有一块城市公共绿地。总建筑群由一栋商务写字楼、数栋商务公寓、三组住宅及一组商业建筑构成，总建筑面积1 054 108 m^2。

Project Planning

Project planning emphasizes regional road axis and urban public green space. Block A is divided into north part that mainly is residence and south part that includes residential area, office apartment area, and business area through urban public green area. Block B is for commercial buildings use. Block C is constructed as noble residential area and open sports park. It plans to make use of cross traffic axis, ring landscape axis and central music fountain square to combine residences and sports park into an organic whole, to make them complement each other.

项目规划

项目规划上强调区域的道路轴线和城市公共的绿地空间。项目A区由城市公共绿地将其分为南地块和北地块。北地块主要为住宅区，南地块则包含住宅区、办公公寓区、商业区三部分。B区为商业建筑用地。C区为高档住宅区及开放式体育公园。规划利用十字形交通轴线、环形景观轴线及中心音乐喷泉广场，将住宅与体育公园结合成有机的整体，相得益彰。

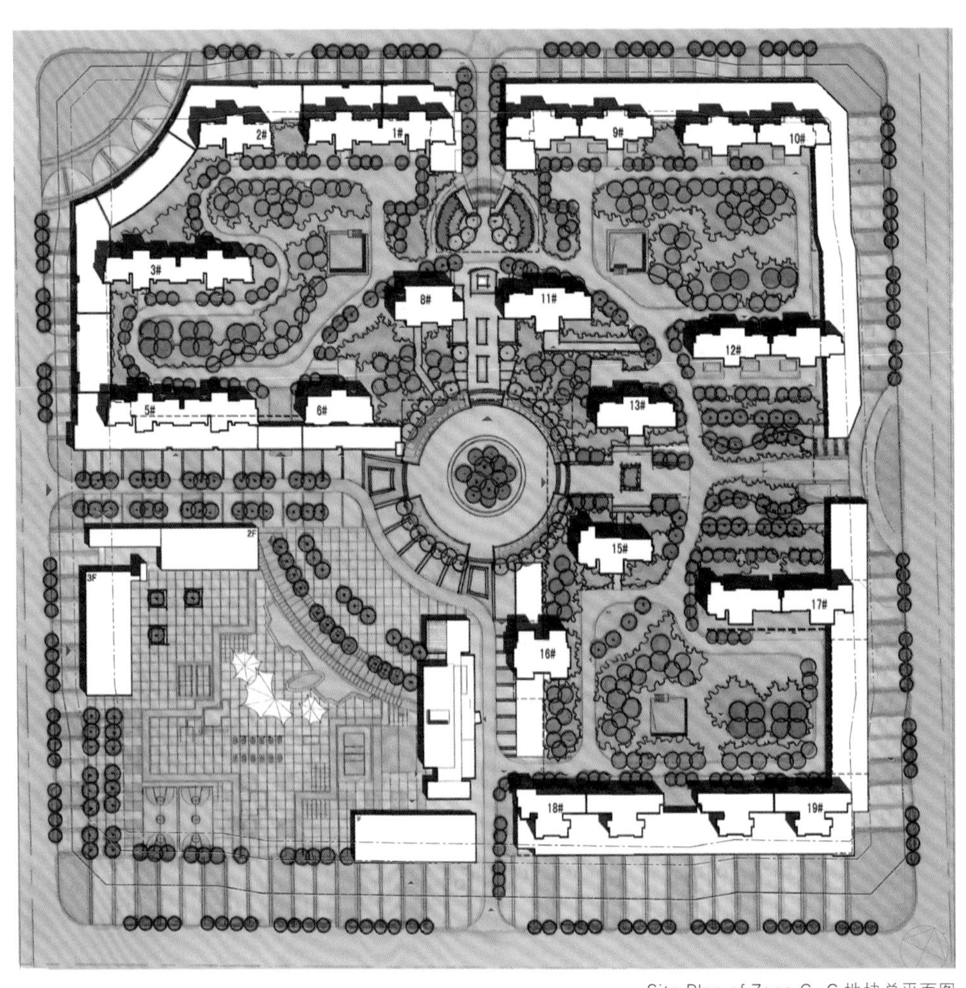

Site Plan of Zone C　C 地块总平面图

➤ Design Concept

The project focuses on openness, which embodies its quality. Taking the central square as the core and the urban public green space as regional center, the project forms a continuous and integral indoor and outdoor open space system through organizing the open space inside the plot and penetrating the open space outside the plot; semi-open community streets lead residents back to their warm home.

➤ 设计理念

项目注重开放性。项目的品质是通过其开放空间得以体现的。以中心广场为核心，城市公共绿地为区域中心，项目通过对地块内开放空间的组织以及对地块外开放空间的渗透，形成连续的、整体的室内外开放空间体系；半开放的社区街道引领居住者回到自己温馨的家。

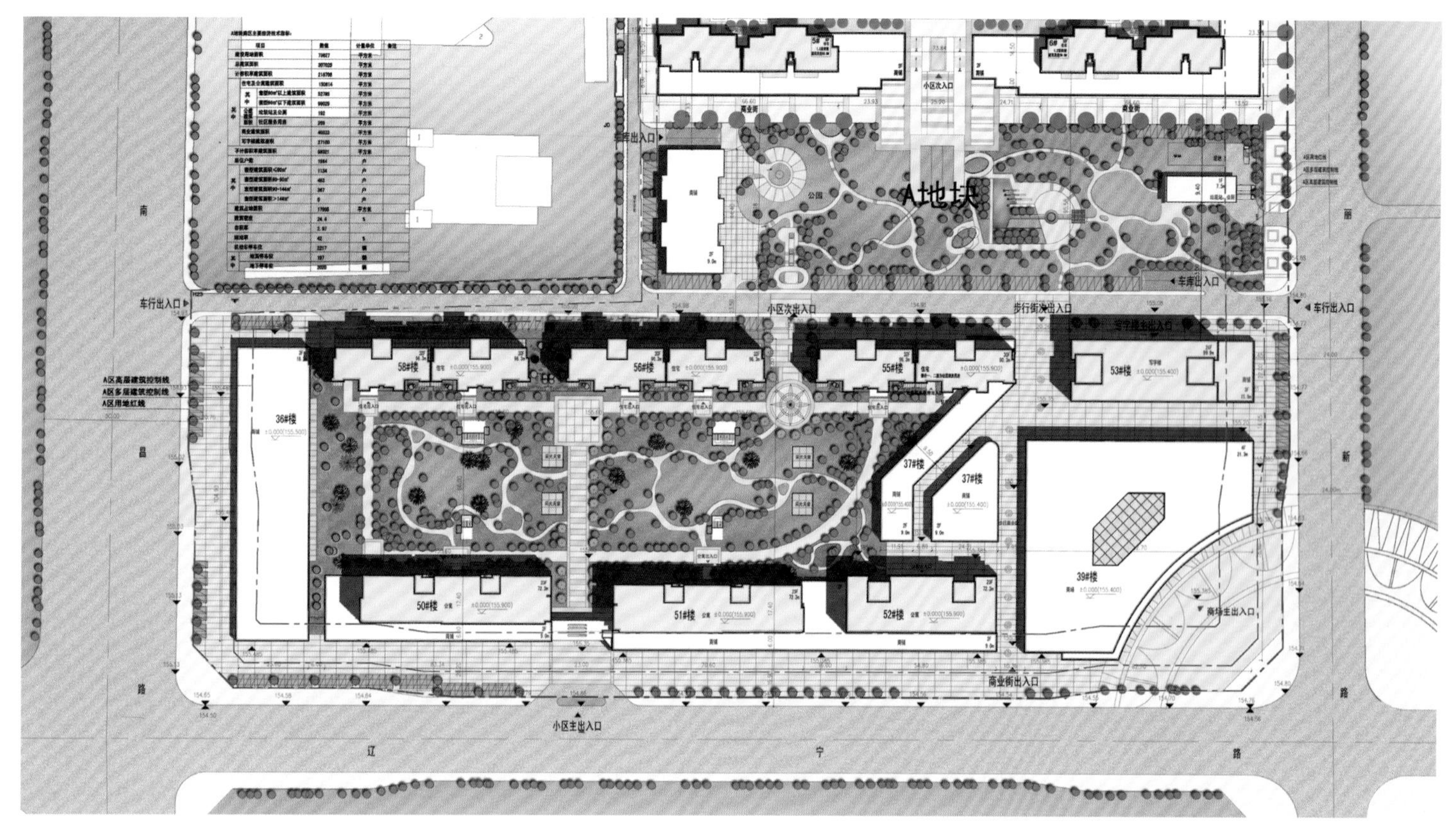

Site Plan of the South Part of Zone A

A 区南地块总平面图

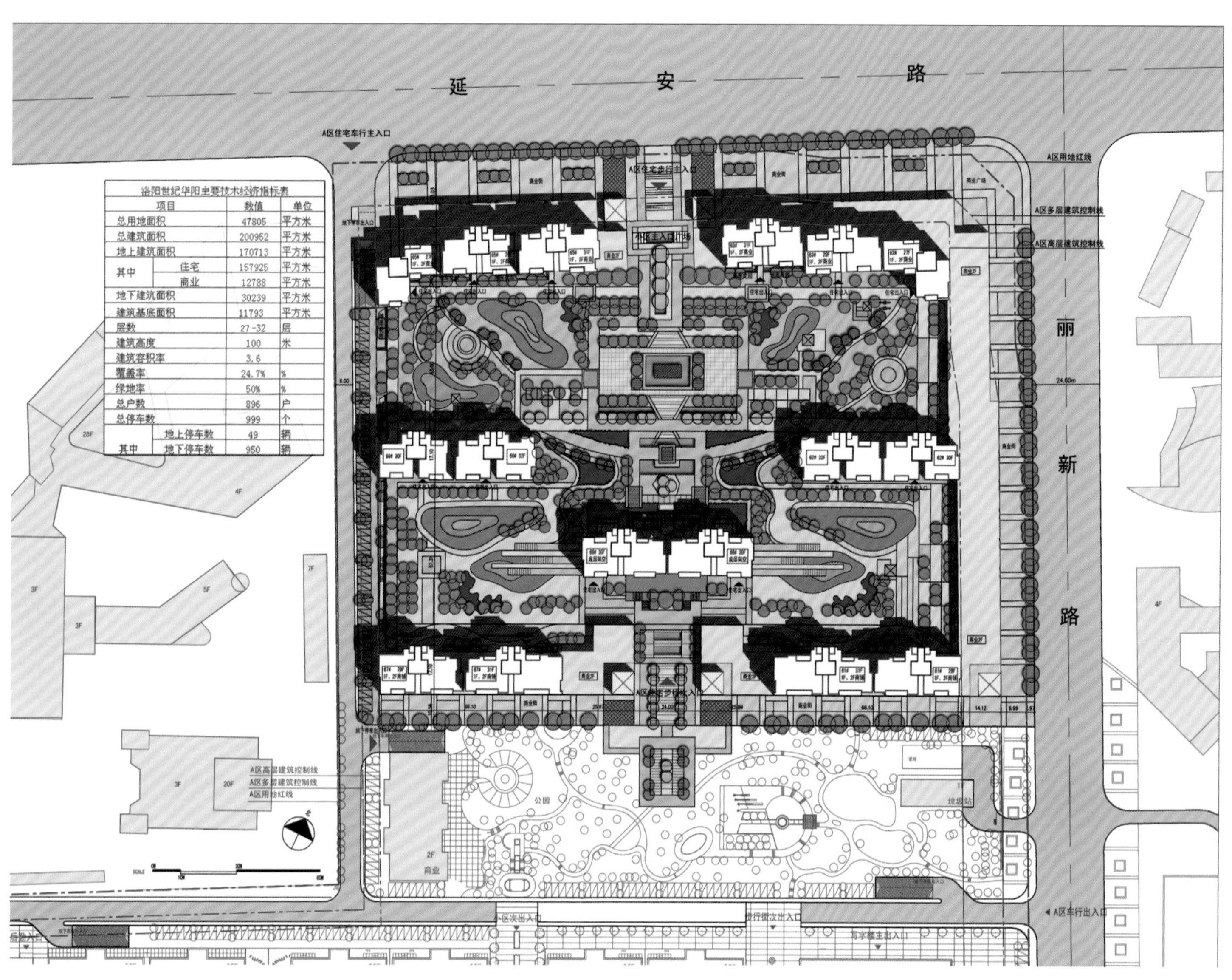

洛阳世纪华阳主要技术经济指标表			
项目		数值	单位
总用地面积		47805	平方米
总建筑面积		200952	平方米
地上建筑面积		170713	平方米
其中	住宅	157925	平方米
	商业	12788	平方米
地下建筑面积		30239	平方米
建筑基底面积		11793	平方米
层数		27-32	层
建筑高度		100	米
建筑容积率		3.6	
覆盖率		24.7%	%
绿地率		50%	%
总户数		896	户
总停车数		999	个
其中	地上停车数	49	辆
	地下停车数	950	辆

Site Plan of the North Part of Zone A

A区北地块总平面图

Architectural Design

The project design of architectural modelling appropriately uses greyish white color tone of traditional Chinese architectural form, and creates a modern community architectural image with Chinese feature through modern architectural materials. Because of the various types of business, there exist various architectural forms. The project always adheres to the keynote of "conciseness, grandness, persistence", and strives to reflect the characteristics and sense of times of modern architecture. The form combination rich in sculptural sense and the concentration on the alternation and strength of simple blocks, give the buildings strong visual impact and unique mark.

建筑设计

项目在建筑造型的设计上适当运用了中国传统建筑形式的灰白色调，利用现代的建筑材料营造出具有中国特色的现代社区建筑形象。由于本区域存在多种业态，因而对应存在着多种建筑形态。项目始终坚持以"简洁、大气、持久"为基调，力求体现现代建筑的特征与时代感。富有雕塑感的形体组合、强调简洁体块的穿插与力量，都使得建筑群具有强烈的视觉冲击力与独特的标志性。

Facade Design

The characterization of facade texture doesn't emphasize the superficial hollow and solid comparison, but simplifies the composition technique. Through the further design of structure, the facade texture tends to be homogenized, unified in a grid system, being structure, surface and as well as facade division, to achieve the maximal uniform between internal function and external performance.

立面设计

立面肌理的刻画，并不强调表面化的虚实对比，而是简化构成手法。通过对结构的深入设计，使立面肌理趋于均质化，统一在一套网格系统中，既是结构，又是表皮，同时又是立面划分，以达到内部功能与外在表现的最大限度的统一。

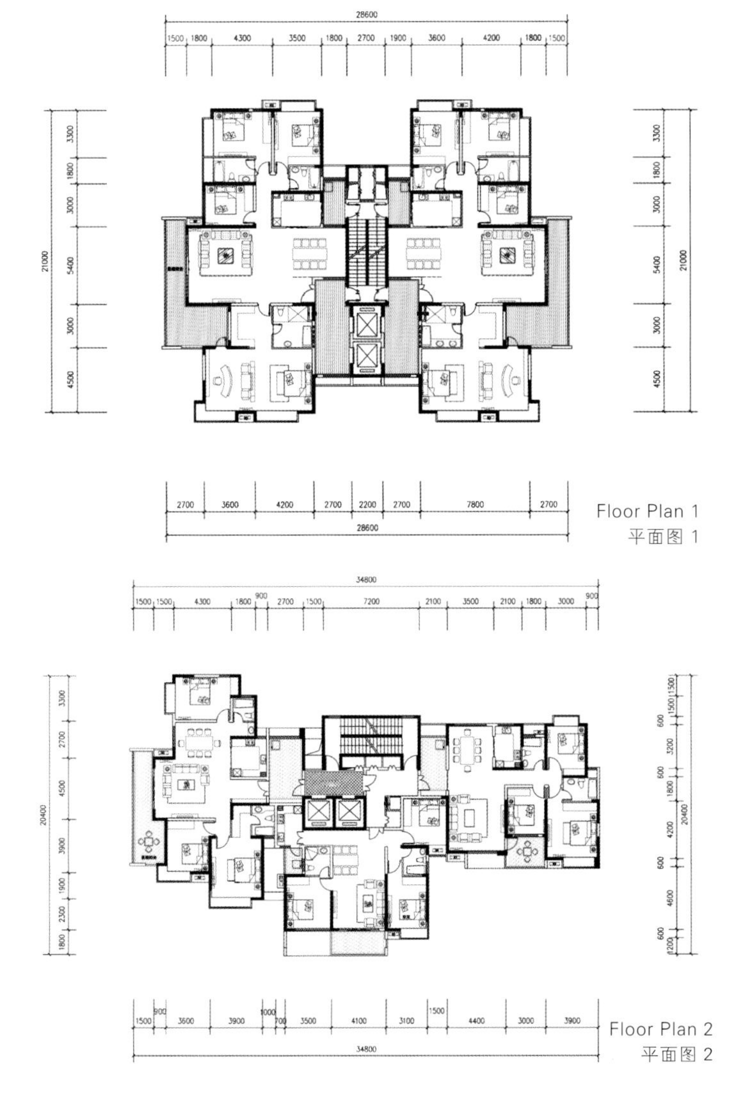

Floor Plan 1
平面图 1

Floor Plan 2
平面图 2

世纪华阳
建业·世纪华阳
ICBC 中国工商银行
秀思美发

世
纪
长申国际广场
长申国际广场
大张集团恢弘巨作
即将精彩呈现
高端百货　国内外一线品牌强势入驻
打造洛阳商业新旗舰

Courtyard Villa, Hefei

合肥·合院别墅

Location: Hefei, Anhui, China
Architectural Design: WSP ARCHITECTS
Land Area: 13,333.4 m^2
Floor Area: 28,408 m^2
Plot Ratio: 1.5

项目地点：中国安徽省合肥市
设计单位：维思平建筑设计（WSP ARCHITECTS）
占地面积：13 333.4 m^2
建筑面积：28 408 m^2
容积率：1.5

FEATURES 项目亮点

The architects decide to use modern techniques to realize the traditional design idea: every villa will be built as a courtyard complete with independent entrance space, front yard, cloister and home garden. All these buildings are presented in a traditional and humanistic sequence.

设计师决定采用现代的表现手法去实现传统的设计理念，将每幢建筑建造成为一个院落，每一个院落都有独立的入口空间、前院、回廊和花园。建筑以一种传统的、人文的序列展示在面前。

KEYWORDS 关键词

Courtyard Unit
院落单元

Harmony Between Human and Nature
天人合一

Modern Techniques
现代手法

➤ Overview

Located in Hefei City, Anhui Province, the site is near to the Twin Phoenix Lake on the east and Heshui Road on the west, enjoying a beautiful environment. Comprising five single-family villas, five semi-detached villas and one large-area high-rise building, it aims to be a high-taste residential community.

➤ 项目概况

项目位于安徽省合肥市，建设用地东临双凤湖，西临合水公路，环境优美。整个别墅区包括五栋独幢的别墅、五栋双拼别墅和一栋大户型高层，力图成为一个高品位的住宅区。

➤ Design Concept

The project design focuses on the harmony between human and nature. In modern fast-paced life, "landscape" and "architecture" are usually regarded as two separate concepts. While in this project, it tries to build a new relationship between them two. Just like human beings to keep harmonious with nature, all the buildings are embraced by landscapes, and a natural, closer and inseparable relationship is thus built between them. When wandering in the villa area, from every corner one will experience the charm of the landscape and architecture, just like standing in the painting.

➤ 设计理念

天人合一：在现代快节奏的生活中，“风景”和“建筑”往往是两个独立的概念。该项目的设计力图改变两者之间的关系。像人与自然的相互包容，每幢建筑都在风景的包围之中，建筑与风景相互交错融合，形成一种自然、朴实、密不可分的关系。漫步于别墅区，在任何角落，都能感受到风景与建筑的魅力，人也仿佛在画中。

Space and Courtyard

The architects decide to use modern techniques to realize the traditional design idea: every villa will be built as a courtyard complete with independent entrance space, front yard, cloister and home garden. All these buildings are presented in a traditional and humanistic sequence. And the introduction of courtyard has combined landscapes with buildings. The courtyard provides more colorful living spaces for family members beyond the architectural spaces. Different courtyards are connected by an axis, along which the public spaces are arranged, to form an integrated neighborhood. With this kind of design, people will not only enjoy happy family life, but also feel satisfied with themselves in neighborhood communication.

空间与院落

设计师决定采用现代的表现手法去实现传统的设计理念，将每幢建筑建造成为一个院落，每一个院落都有独立的入口空间、前院、回廊和花园。建筑以一种传统的、人文的序列展示在面前。院落的导入实现了风景与建筑的统一。院落为家庭提供了比建筑更为丰富的生活场景。院落成为了构成邻里生活的一个细胞，院落的并置串联形成了院落群轴。在这条构图关系的轴线上，公共生活得以展开，居住在其中的人们不仅能享受家庭的天伦之乐，也能从邻里交往中获得社会属性的满足感。

Single Building Design

According to the master plan, houses of different types are built in different areas. The semi-detached houses, with a floor area of about 380 m^2, are built at the main entrance, forming the "peripheral wall" for the community. And the large-area townhouses (about 1,000 m^2) are built by the lake in the southeast to enjoy great lake views. At the south end, 500 m^2 townhouses are designed to have ample sunlight and beautiful views of the wood. And about 400 m^2 sky villas are built on the lakeside in the northeast to overlook the beautiful Twin phoenix Lake.

建筑单体设计

根据别墅区的总体规划，不同的位置设计了不同的户型。主入口处为双拼别墅，每户面积在380 m^2左右，形成了小区的“墙”。东南临湖设计成1 000 m^2左右的联排大别墅，充分享受沿湖风景。南端为500 m^2的联排别墅，住户可享受南端充足的阳光以及美丽的树林。东北临湖布置400 m^2左右的空中别墅，鸟瞰双凤湖的旖旎风光。

British Style
英式风格

Noble and Elegant

高贵典雅

Beautiful and Exquisite

美观精致

Rustic and Dignified

古朴厚重

Chengdu Poly Central
成都保利中央峰景

Location: Chengdu, Sichuan, China
Construction Unit: Poly Southern (Sichuan) Investment Development Co., Ltd.
Architectural Design: Huayang International Design Group
Cooperative Design: Sichuan Zhongheng Architectural Design Co., Ltd. (Construction Drawing Design)
Land Area: 94,993.64 m^2
Gross Floor Area: 286,652.18 m^2
Plot Ratio: 2.37
Awards: the 13th Chinese Real Estate Development Annual Meeting·2013 Chinese City New Landmark Award (Chengdu)

项目地点：中国四川省成都市
建筑单位：保利南方（四川）投资开发有限公司
设计单位：华阳国际设计集团
合作设计单位：四川众恒建筑设计有限责任公司（施工图设计）
占地面积：94 993.64 m^2
总建筑面积：286 652.18 m^2
容积率：2.37
获奖情况：第十三届中国房地产发展年会·2013中国城市新地标（成都）

KEYWORDS 关键词

Beautiful and Grand
美观大气

Low-density Residential Area
低密度居住区

British Style
英伦风情

FEATURES 项目亮点

As Poly's first project in Taoyuan New Town, Poly Central has introduced Dayi Ecological Park views and a suburban residential atmosphere to create an ecological urban garden in pure British style.

作为桃园新城的开篇之作，项目以大邑生态公园资源的引入、郊外住区场所感的塑造，勾勒出一个具有纯正的英伦原乡风情的城市生态后花园。

Overview

Poly Southern plans to make suburb large community construction at Dayi Taoyuan New Town, Chengdu, including commerce, residence, office, culture & sport leisure, hotel and other urban functions. Poly Central, which is located at the core of this new town, is the first construction project, consisting of south and north plots, planned and designed by Huayang International. The north area of phase I includes mainly three types of products, i.e., club and townhouse, garden house and high-rise residence.

项目概况

保利南方拟在成都大邑桃园新城进行市郊大社区建设，规划包括商业、住宅、办公、文体休闲和酒店等多种城市功能。位于新城核心位置的保利中央峰景是其首个建设项目，分南北两个地块，由华阳国际统一规划设计。一期建设的北区主要由会所及联排别墅、花园洋房、高层住宅三种产品组成。

Planning Objective

Although this project is located at the core area of the new town and Chengdu and Dayi will be connected by rail transit and fast channel in the future, the present situation is that it is still of certain distance from the mature urban area. Around this land, there is only an expressway leading to the Xiling Snow Mountain and there are no other residential areas or supporting urban functions, so "suburbanization" is its key character. Since the owners expect a kind of "prosperity and countrified garden coexisting" life style, the creation of sense of place and complete supporting function planning become the challenge and mission of this project design.

规划目标

虽然项目位于新区的核心位置，未来可通过轨道交通及快速通道便捷地连通成都大邑双城，但现在项目离成熟的市区仍然有一定的空间距离。用地周围除了一条通往西岭雪山的快速路，既没有其他住宅区，也没有配套的城市功能，"郊区化"是其重要特征。而业主希望在此打造一种"进则繁华，退则田园"的生活方式，因此场所感的营造和完备的配套功能规划成为本项目设计面临的挑战与使命。

Site Plan
总平面图

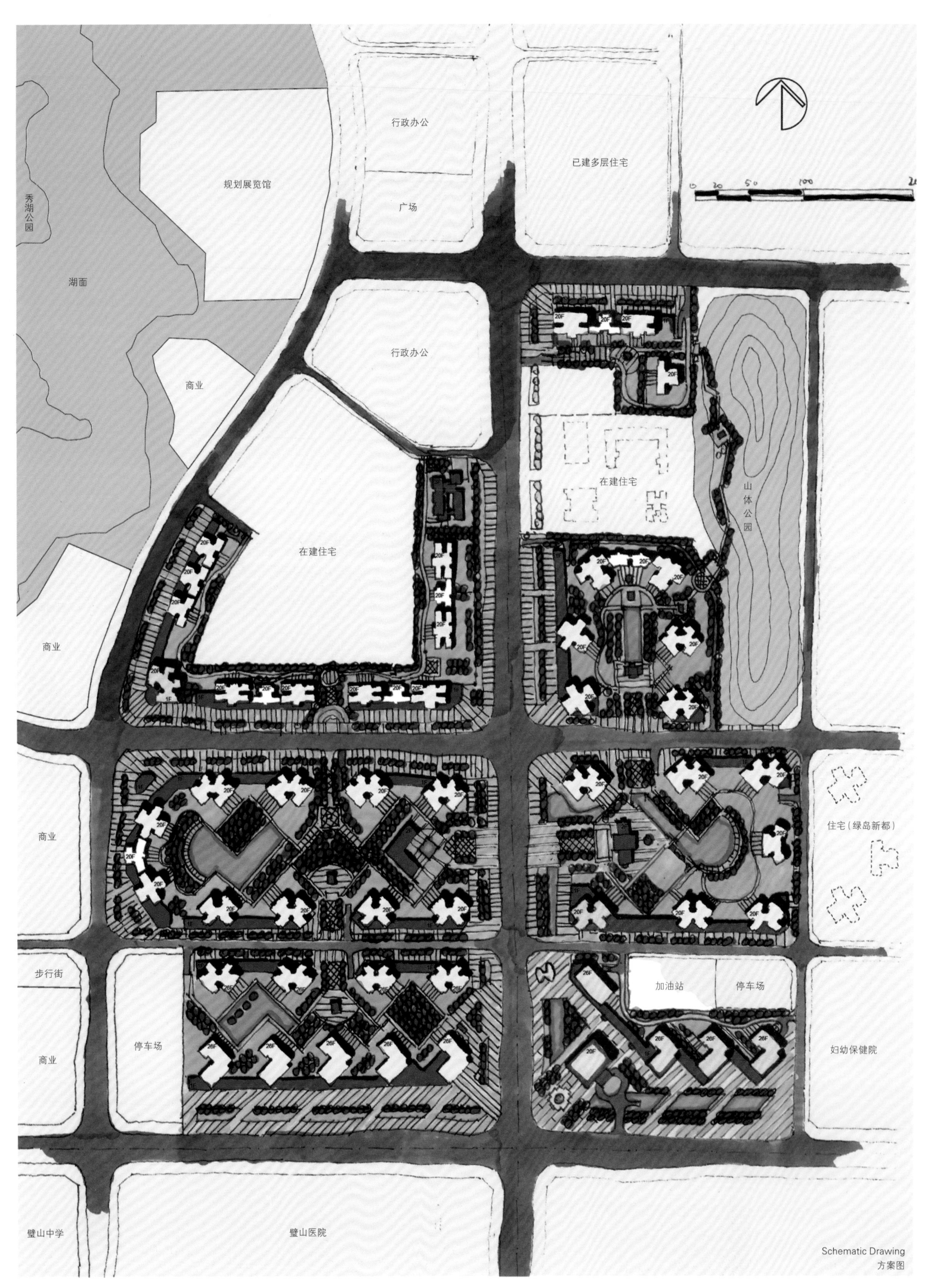

Schematic Drawing
方案图

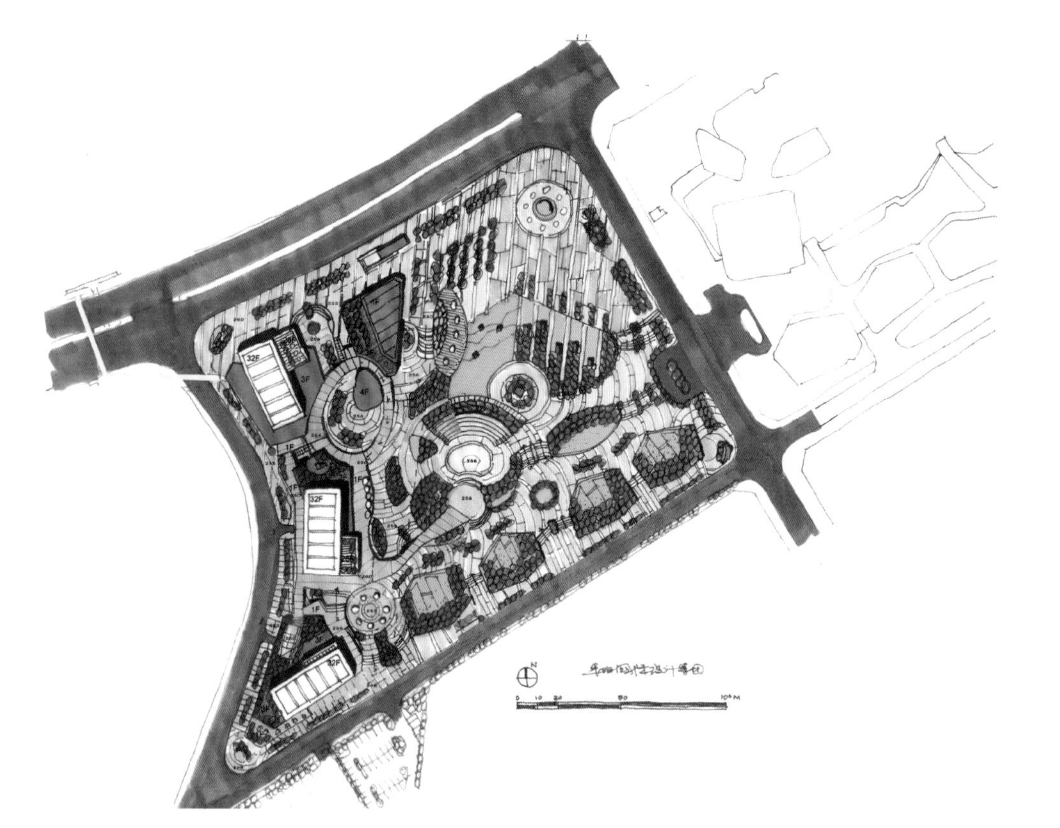

➤ Project Positioning

To start large-scale community construction at this "nothing at all" place, the first problem to be solved is the positioning of this project. Relative to those residential areas at the old town, the planning and construction of the municipal parks on the west and south sides of this project offer the chance of building "park residential area" and "suburb slow life experience" for this project. The designers position this project as a high-end low density ecological residencial area at the center of Taoyuan International New Town according to the character of the park.

➤ 项目定位

在这个被认为"什么地方都不是"的位置开始大型社区建设，首先要解决的就是项目的定位问题。相对于那些旧城区的住宅区而言，项目西侧和南侧规划建设的市政公园为其提供了打造"公园化住区"和"郊外慢生活体验"的机会。设计师通过挖掘公园特点，将项目定位为桃园国际新城中央的高端低密度生态住宅。

Elevation 1
立面图 1

Elevation 2
立面图 2

To balance the internal plot ratio, the design takes "park-like entrance, high west and low east" as the planning guideline. The high-rise residential buildings are placed at the west plot, while the townhouse and foreign-style house are placed at the plot near the park, and the spatial picture of a British town is fully displayed at the entrance. It includes lake surface, wooden bridge, clubs, Tudor style business street and Victorian style business street and other British town elements, as well as the Victorian style townhouse. British style building size and human landscape elements are integrated to build a rich country environment, and meet the necessary community life requirements, building a delicate and elegant, beautiful and grand community with rich architectural elements and strong cultural atmosphere, and providing a clear symbolic community impression for the people coming into the community.

Townhouses Elevation 1
联排立面图 1

Townhouses Elevation 2
联排立面图 2

为了平衡内部容积率，设计以“公园化入口，西高东低”作为规划思路。高层住宅集中布置在西侧地块，联排别墅和洋房布置在靠近公园的位置，并试图在项目入口区完整地展示一个英伦小镇的空间图景。这里包括湖面、木桥、会所、都铎风情商业街与维多利亚风情商业街等多种英伦小镇元素，也包括了维多利亚风格的联排别墅。英伦风情的建筑体量和人文景观元素共同营造出浓郁的田园特征场所感，基本满足了社区生活必要的功能配套，也营造出精致典雅、美观大气、建筑元素丰富、富有浓厚文化气息的社区，为所有走进社区的人提供了一个清晰的符号化的区位印象。

Chief Park City (Phase I), Xi'an
西安建秦锦绣天下（一期）

Location: Xi'an, Shaanxi, China
Architectural Design: BDCL Design International Co., Ltd.
Gross Land Area: 129,800 m^2
Gross Floor Area: 157,100 m^2
Plot Ratio: 1.21

项目地点：中国陕西省西安市
设计单位：博德西奥（BDCL）国际建筑设计有限公司
总占地面积：129 800 m^2
总建筑面积：157 100 m^2
容积率：1.21

FEATURES 项目亮点

The buildings adopt the three-segment design. The interweaving and interlocking of different blocks add a few of vibrant elements to the originally serious facade.

建筑采用三段式设计，但又通过一些体块的穿插、咬合关系打破了通常的三段切分方式，给稳重的立面增添了些许活跃的元素。

KEYWORDS 关键词

British Style
英伦风格

Elegant and Dignified
优雅端庄

Denotative Image
外延形象

Overview

Located in Weiyang District of the northern Xi'an City, belonging to Chanba Ecological Base planning scope, the project is near to Weiyang Lake University Town on the north, and Chanba River on the east, overlooking Xi'an new administrative center on the west. It is envisioned to be a noble benchmark community image with low-density, elevator-equipped garden apartments, superior to the others in this area for its attractive architectural form and high quality.

项目概况

项目位于西安市北未央区内，在浐灞生态基地规划范围内。北临未央湖大学城，东依浐灞河，西望西安新的行政中心。目标定位为打造低密度电梯花园洋房产品，营造高品质标杆产品，以产品形态及品质树立高端大盘住区形象。

Overall Layout

The planning layout is characterized by "one ring" "two belts" "one center" and "one gateway". Curved road brings a dynamic and rich space experience, and two landscape axes provide changeful public green spaces. At the cross of these two axes, the landscape center of the community is decorated by different elements to be a comfortable space for neighborhood communication. At the main entrance, the sales center, business, as well as the gate and peripheral wall are integrated to form a kind of high-end and comfortable entrance atmosphere. The entrance space has served as the gateway of the community and the starting point of the south-north landscape axis.

总体布局

规划布局以"一环""两带""一心""一门户"为主要特点。曲线的道路为小区营造了动态丰富的空间体验；两条主要景观轴线提供了宜人且富有变化的公共绿色空间，两轴交会处则是小区的景观中心，通过景观元素营造良好的视觉中心，成为宜人的邻里交往空间；主入口结合销售中心、商业、大门及围墙营造了高端宜人的入口氛围，成为小区的门户和南北景观轴的起点。

18栋2单元

➤ Facade Style

The buildings adopt simple and modern British style and three-segment design. The interweaving and interlocking of different blocks add a few of vibrant elements to the originally serious facade.

➤ 立面风格

建筑采用简约现代版的英伦风格和三段式设计，但又通过一些体块的穿插、咬合关系打破了通常的三段切分方式，给稳重的立面增添了些许活跃的元素。

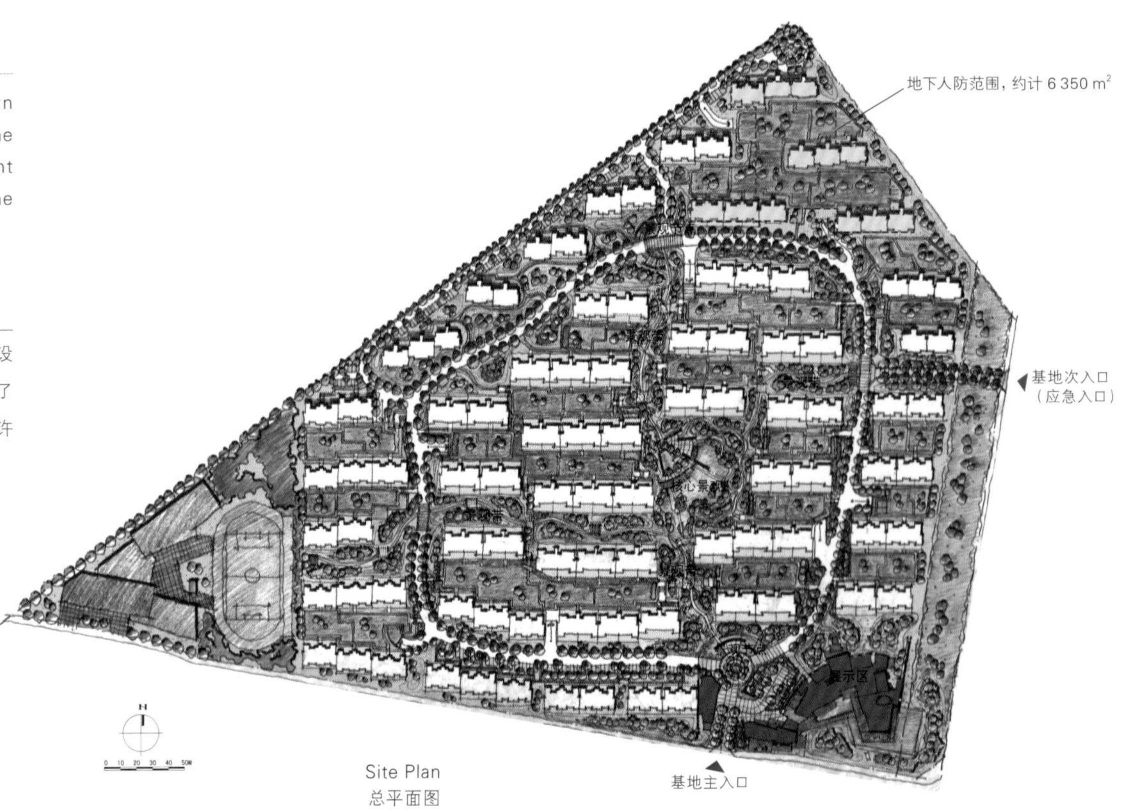

Site Plan
总平面图

Facade Material

The buildings look elegant and dignified with wall tiles and stones as the main facade materials. Some additional metal and glass components add vitality to the buildings. Red tiles define the tone of surface and split tiles of other colors are interspersed to highlight the texture of the buildings. The foundation part uses gray granite to emphasize its approachable scale and high quality. Part of the building top is painted in light color to make the steady-going building breathable.

立面材料

建筑立面的主要材料为面砖与石材，优雅端庄；局部穿插的金属与玻璃又为建筑增加了活力。以暖红色温暖的面砖为主，同时局部穿插同色系不同色相的劈开砖，以提升建筑的质感。基座采用灰色花岗岩，以提升近人尺度及品质感。顶部局部采用浅色涂料，给沉稳的建筑以局部的透气感。

House Type Design

The apartments are designed with different house types. 80% of them are units with two bedrooms or three bedrooms. The apartments on the ground floor are accessible through the gardens, and the duplexes on the top are designed with double-height living room and terrace, enjoying the space experience of villas and promoting their added value. The apartments above the second floor employ home gardens, terraces or balconies to create a garden-like environment. All these apartments are designed with independent porch, flowing living room and dining room to extend the space experience.

户型设计

住宅产品丰富，主力户型为两房两厅和三房两厅，占开发面积比例和总套数比例均在80%左右。首层户型从花园直接入户，顶层复式户型客厅挑空，设露台，为首层、顶层户型营造别墅感，提升附加值。二层以上户型尽可能营造花园感（如利用入户花园、露台、阳台等方式）。入户皆设独立玄关，起居厅与餐厅空间流动，延伸了空间体验。

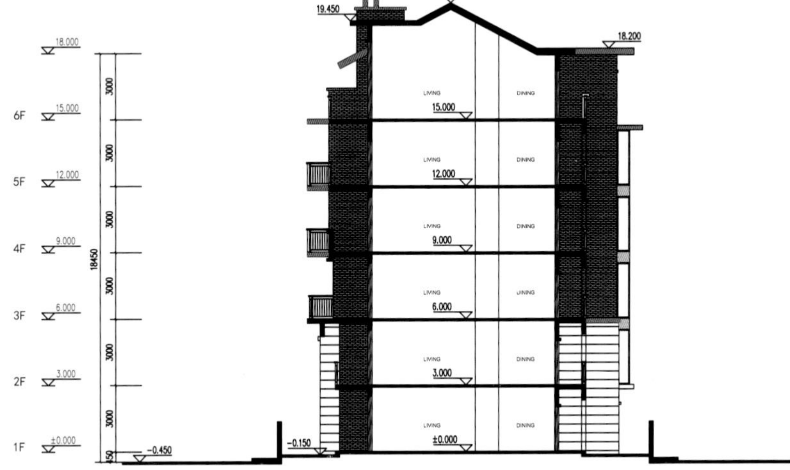

4# 1-1 Section 4# 楼 1-1 剖面图

➤ Clubhouse Design

The sales center, business, gate and the surrounding landscapes are under integrated design to save cost and achieve a shocking effect: the coherent sales area will impress people a lot.

Based on the landmark tower axis, the buildings are well organized to be opposite, staggered or twisted, with strong connection between themselves. The tower has become the focus and key element for this area.

It employs courtyard, roof overhung and landmark tower to highlight the denotative image of the buildings. Because the land for building is limited, to meet the sales requirements, it needs to create an impressive selling atmosphere. Thus the designers employ courtyard, roof overhang and landmark tower to expand building structures and achieve the best exhibition effect with lower cost.

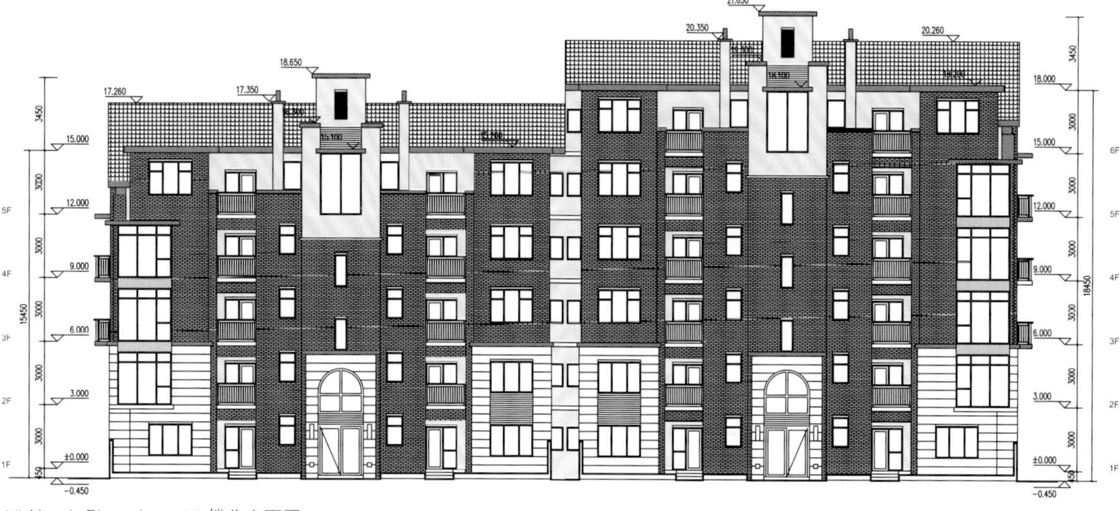

4# North Elevation 4# 楼北立面图

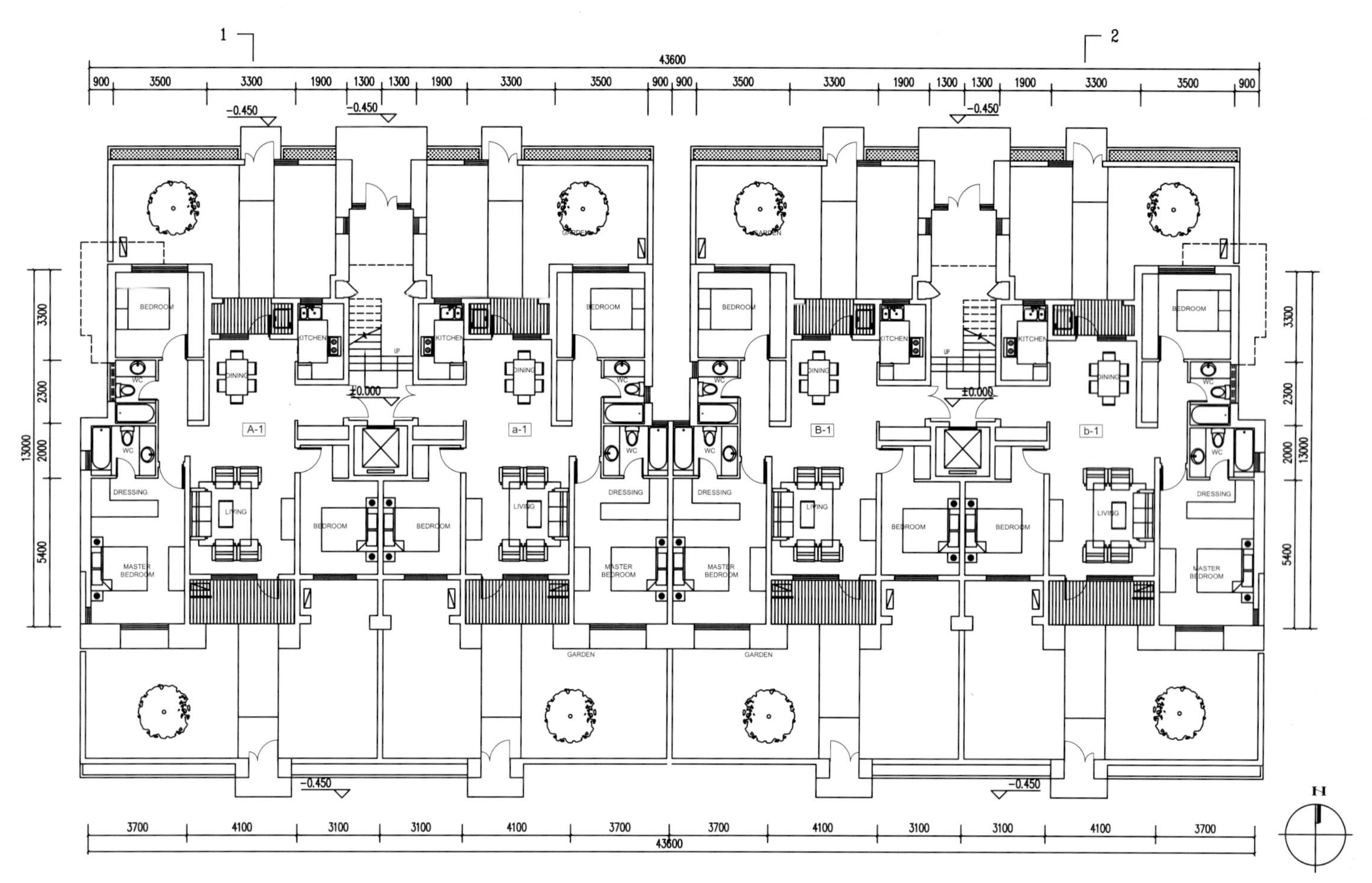

First Floor Plan
首层平面图

➤ 会所设计

将销售中心、商业、大门及周围景观当作一个整体来设计，以低成本的代价取得震撼的效果，令销售动线的整体印象非常连贯。

通过标志塔轴线组织建筑形体关系。为形成有张力的组织关系，几栋建筑采用了一些对位、错动和扭转等动感的构图元素，通过强有力的轴线关系将它们组织在一起。标志塔成为整组建筑的构图焦点和关键元素。

项目运用院墙、挑檐、标志塔等构筑物，扩大建筑的外延形象。由于建筑用地比较有限，但为满足销售造势的需求，需要营造气势恢宏的卖场形象。因此设计师借用院墙、挑檐、标志塔等构筑物，延展建筑的形象范围，从而用较低的成本获得最佳的展示效果。

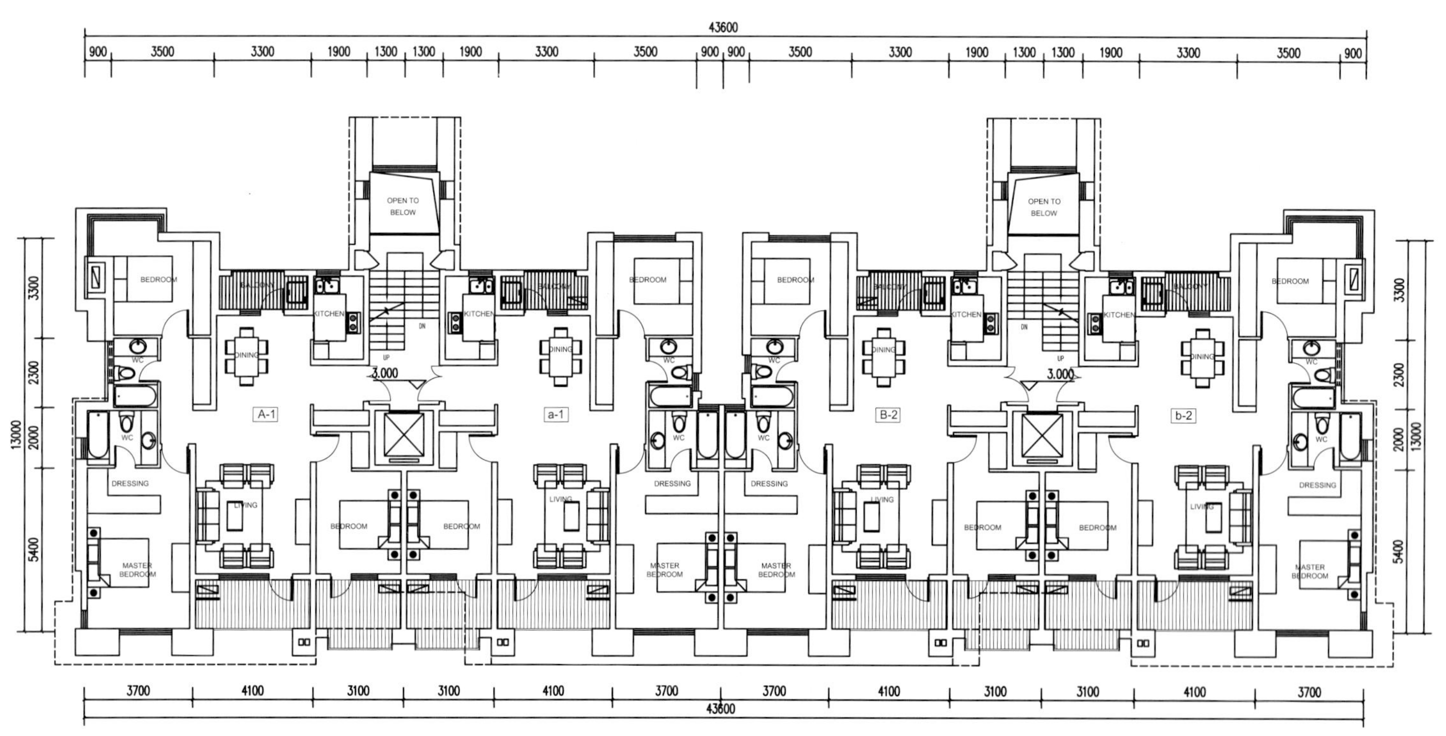

Second Floor Plan
二层平面图

Tianjin Xinbang Ruijing (Plot 14)

天津信邦瑞景（14 号地块）

Location: Beichen District, Tianjin, China
Architectural Design: Tianjin University Research Institute of Architectural Design & Urban Planning
Floor Area: 107,170 m^2
Site Area: 82,600 m^2
Plot Ratio: 1.31

项目地点：中国天津市北辰区
设计单位：天津大学建筑设计研究院
建筑面积：107 170 m^2
占地面积：82 600 m^2
容积率：1.31

FEATURES 项目亮点

The overall layout of the project foucuses on multi-storey residences. Seventeen multi-storey residences are built in the center of the site to ensure a high environmental quality. And on the north and east sides are five high-rise residences and one high-rise apartment building. The overall layout enjoys a clear partition.

整体布局以多层住宅为主，在用地的核心区域布置了 17 栋多层住宅，使小区具有较高的环境质量。沿用地北侧及东侧布置 5 栋高层住宅及一栋高层公寓，总体布局分区明确。

KEYWORDS 关键词

Beautiful Outline
美丽的轮廓线

Clear Partition
分区明确

Visual Corridor
视觉通廊

➤ Overview

The project comprises multi-storey residences (6-storey and 18.8 m high), high-rise residences (16-storey and 47.7 m high) and public buildings (3-storey and 13.8 m high), with a gross floor area of 107,170 m^2.

➤ 项目概况

项目由多层住宅（6层，建筑高度18.8 m）、高层住宅（16层，建筑高度47.7 m）及公建（3层，建筑高度13.8 m）组成，建筑总面积达107 170 m^2。

➤ Planning Layout

The main exit is set at the south end of Longyan Road, along which stores are arranged, and it is close to the nearest metro station and bus station. These will not only increase the commercial value of the community but also form a good urban interface. The overall layout of the project foucuses on multi-storey residences. Seventeen multi-storey residences are built in the center of the site to ensure a high environmental quality. And on the north and east sides are five high-rise residences and one high-rise apartment building. The overall layout enjoys a clear partition in accordance with this clear planning, the high-rises surround the multi-storey buildings to form a beautiful outline as well as flexible and dynamic internal spaces.

➤ 规划布局

小区的主出口设置在南端龙岩路段，商铺沿龙岩路段布局，邻近地铁站和公交站，既提高了小区的商业价值，又有助于形成良好的城市界面。整体布局以多层住宅为主，在用地的核心区域布置了17栋多层住宅，使小区具有较高的环境质量。沿用地北侧及东侧布置5栋高层住宅及一栋高层公寓，总体布局分区明确。高层对多层的环抱形成美丽的建筑轮廓线和完整的区域内空间，灵活而富于动感。

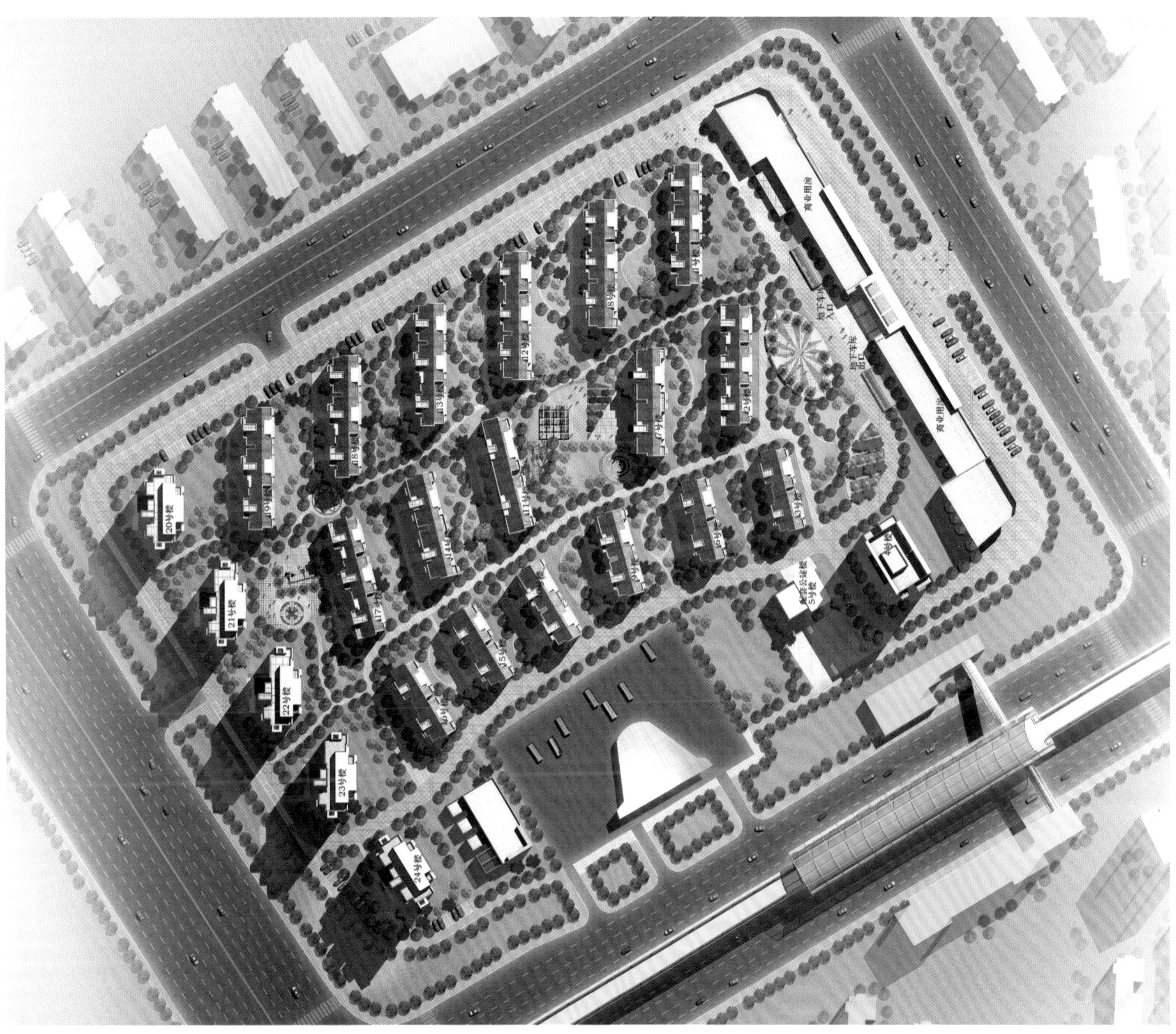

Site Plan
总平面图

➤ Landscape Design

At the main entrance in the south, in the center of the site, and between the northern high-rises and the multi-storey residences, three green spaces are designed to blend with the Liuyuan Nursery Garden on the north both visually and spatially. Together with the clear arrangement of the buildings, there forms two landscape axes from south to north. And people can take a view of these green spaces across the gable walls and peripheral walls. Thus a visual corridor is established to optimize the environment of the community.

➤ 景观设计

在南侧主入口附近、用地中部及北侧高层与多层住宅之间形成了三片集中绿地，它们与用地北侧的刘园苗圃在视线和空间上相互渗透、贯穿，结合住宅单体布局，形成南北向布局的两条主要景观轴线。住宅山墙和檐墙大部分都能望见集中公共绿化，形成良好的视觉通廊，优化了小区环境。

Unit A (Standard Unit 8 Mirror Image)
A 单元（标准单元八镜像）

Unit B (Standard Unit 9)
B 单元（标准单元九）

Unit B (Standard Unit 8)
B 单元（标准单元八）

First Floor Plan 一层组合平面图

South Elevation 南立面图

Building 2 & 17 (Multi-storey Residence) 2、17 号楼（多层住宅）

North Elevation 北立面图

South Elevation 南立面图

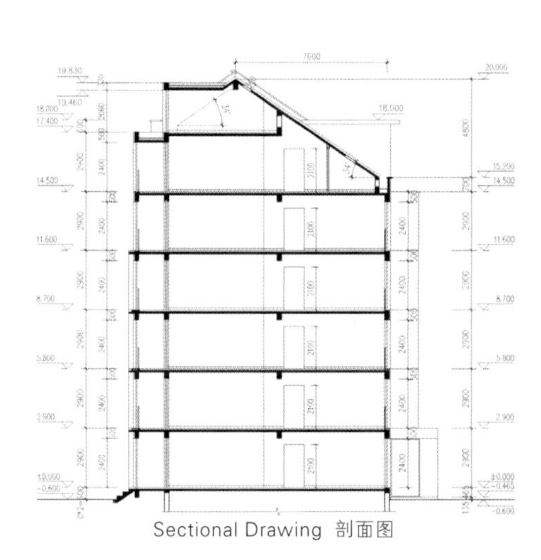

Sectional Drawing 剖面图

Building 1, 13, 18 & 19 (Multi-storey Residence) 1、13、18、19 号楼(多层住宅)

North Elevation 北立面图

End Elevation 侧立面图

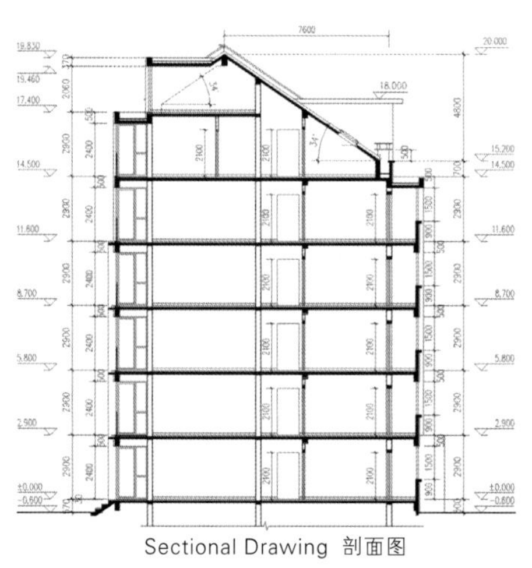

Sectional Drawing 剖面图

Building 2 & 17 (Multi-storey Residence) 2、17 号楼(多层住宅)

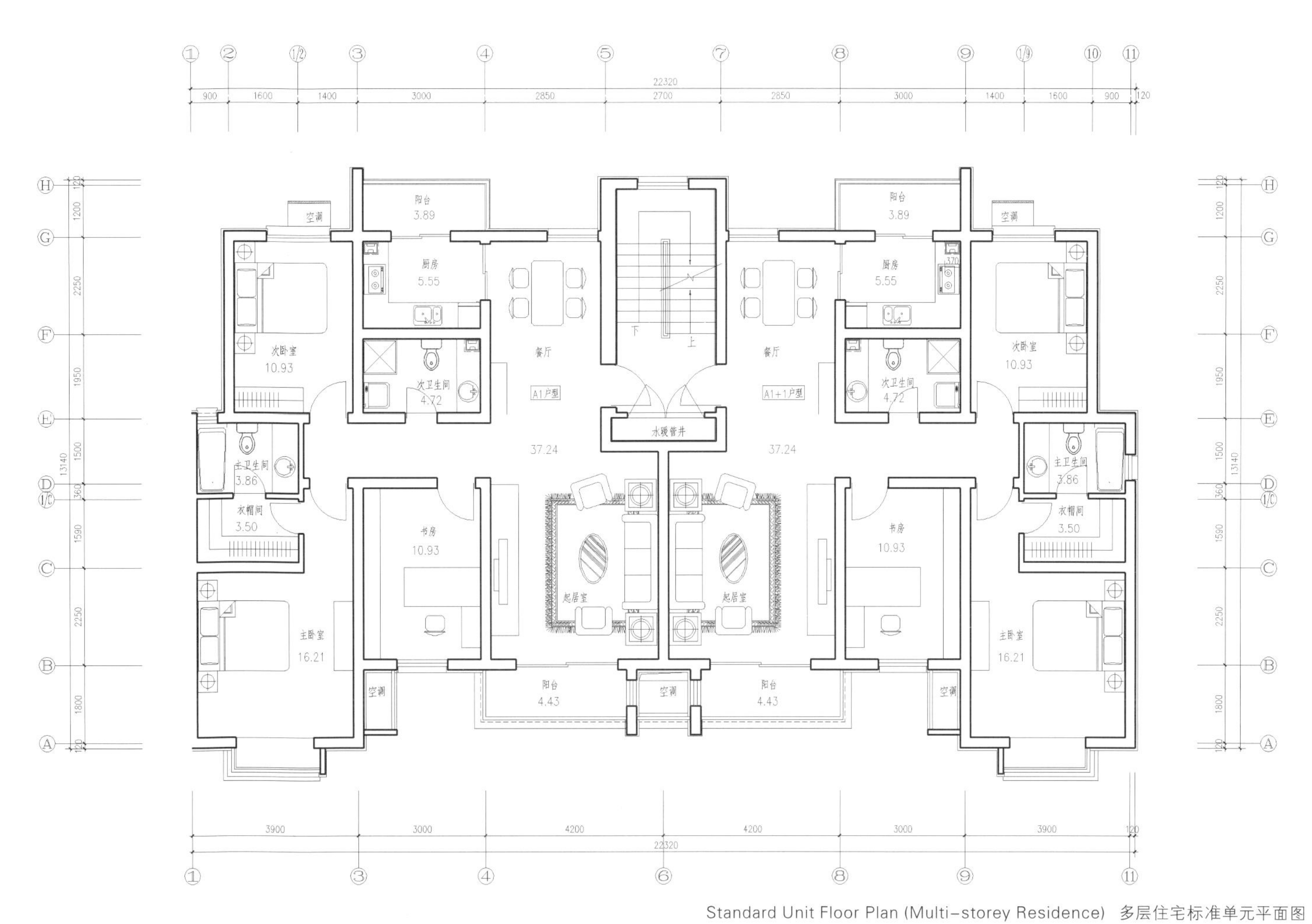

Standard Unit Floor Plan (Multi-storey Residence) 多层住宅标准单元平面图

American Style
美式风格

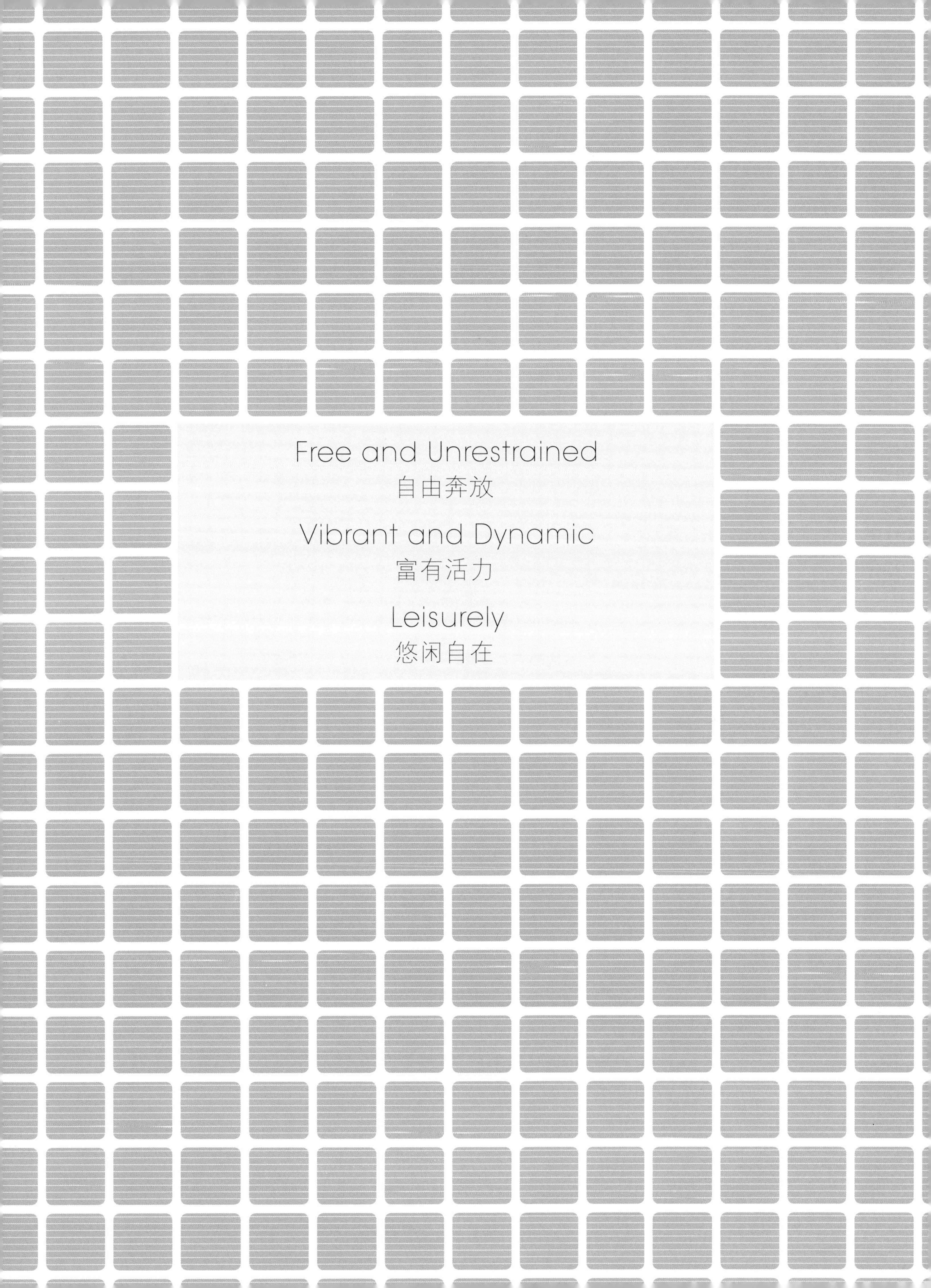
Free and Unrestrained
自由奔放
Vibrant and Dynamic
富有活力
Leisurely
悠闲自在

Niutuo Hot Spring Peacock City

牛驼·温泉孔雀城

Location: Langfang, Hebei, China
Owner: CFLD
Architectural Design: Concord Design Group
Gross Floor Area: 470,000 m^2
Over ground Floor Area: 380,000 m^2
Building Coverage Ratio: 27.0%
Plot Ratio: 1.30
Green Coverage Ratio: 36.1%

项目地点：中国河北省廊坊市
业主：华夏幸福基业
设计单位：西迪国际/CDG国际设计机构
总建筑面积：470 000 m^2
地上建筑面积：380 000 m^2
建筑覆盖率：27.0%
容积率：1.30
绿化覆盖率：36.1%

KEYWORDS 关键词

Prairie Style
草原风格

Chinese Elements
中式元素

Hotel-grade Landscape
酒店级景观

FEATURES 项目亮点

High quality materials, combining with the prairie style expressing the holiday space and the traditional Chinese elements, together with the exquisite hotel-grade landscape system, have perfectly presented the taste of this high-end hot spring resort.

优质材料结合草原风格对度假空间的表现力及传统中式元素的点睛之笔，配以酒店级精致景观体系，完美呈现出“牛驼•温泉孔雀城”的高端温泉度假休闲产品的气质。

➤ Architectural Design

The architects take advantage of the hot spring, and the spirit and regional advantage of CFLD, to highlight the specialized design of the hot spring resort. High quality materials, combining with the prairie style expressing the holiday space and the traditional Chinese elements, together with the exquisite hotel-grade landscape system, have perfectly presented the taste of this high-end hot spring resort.

➤ 建筑设计

在设计中建筑师利用项目得天独厚的温泉资源和华夏地产品牌精神及区域优势，突显温泉主题的升级型度假产品专项设计。优质材料结合草原风格对度假空间的表现力及传统中式元素的点睛之笔，配以酒店级精致景观体系，完美呈现出“牛驼・温泉孔雀城”的高端温泉度假休闲产品的气质。

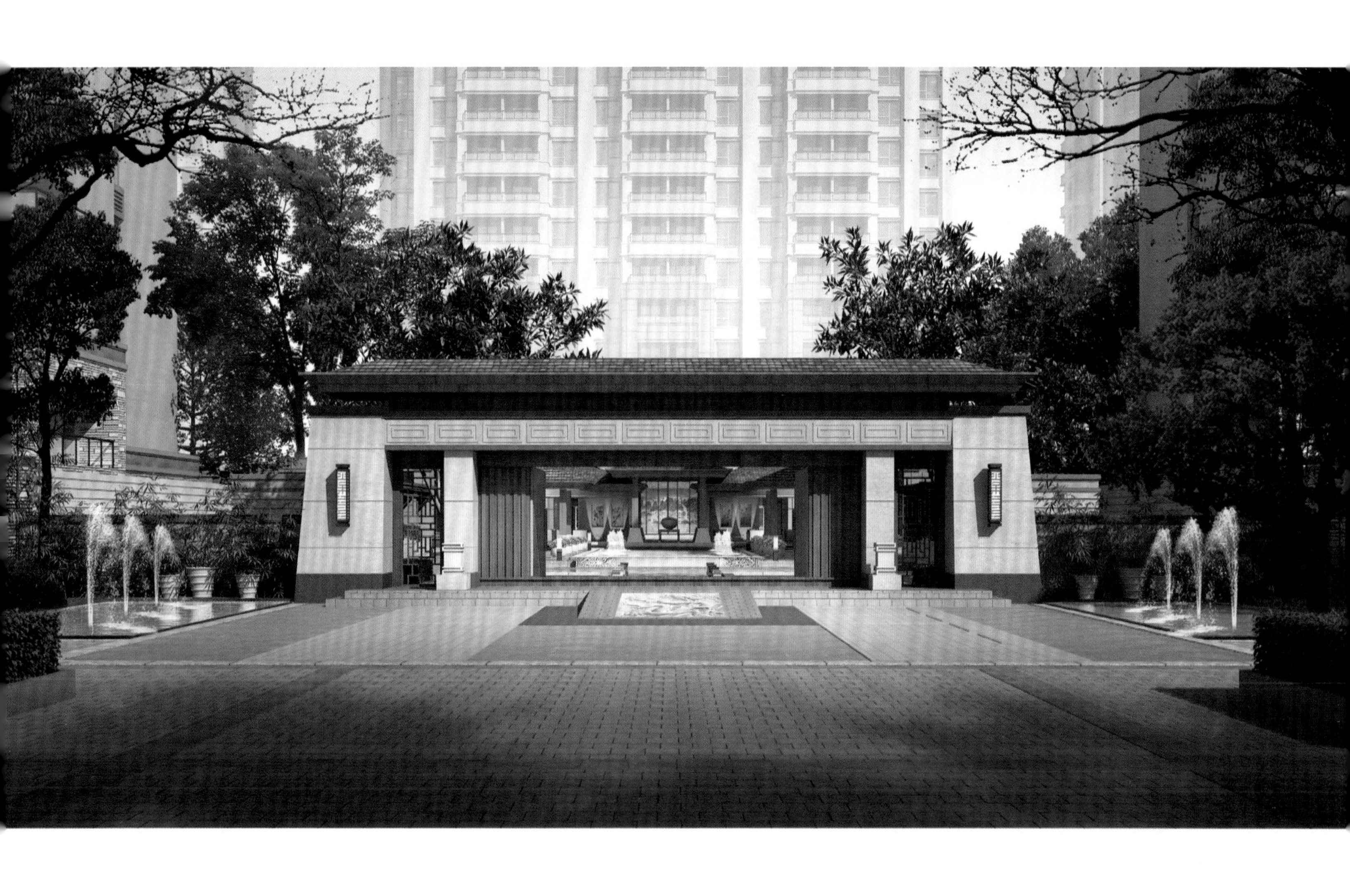

Site Plan
总平面图

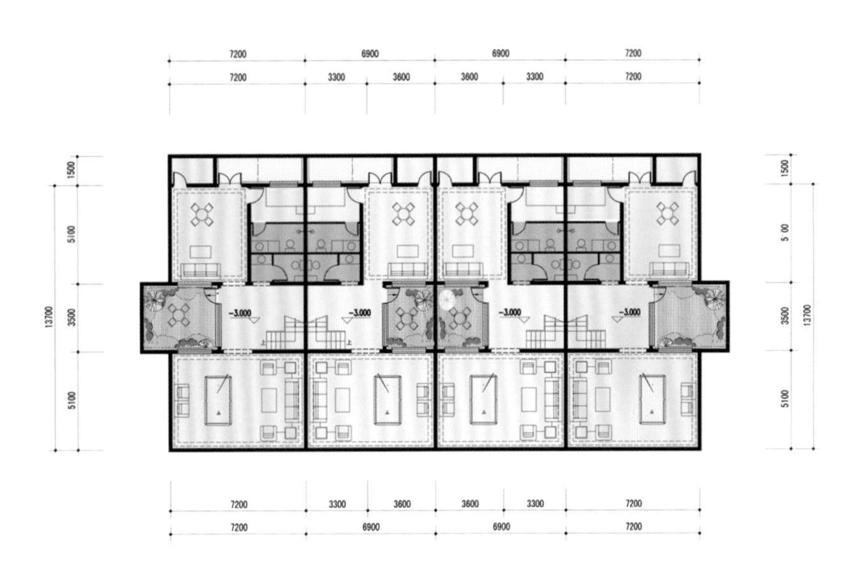

Plan for Basement One Floor
地下一层平面图

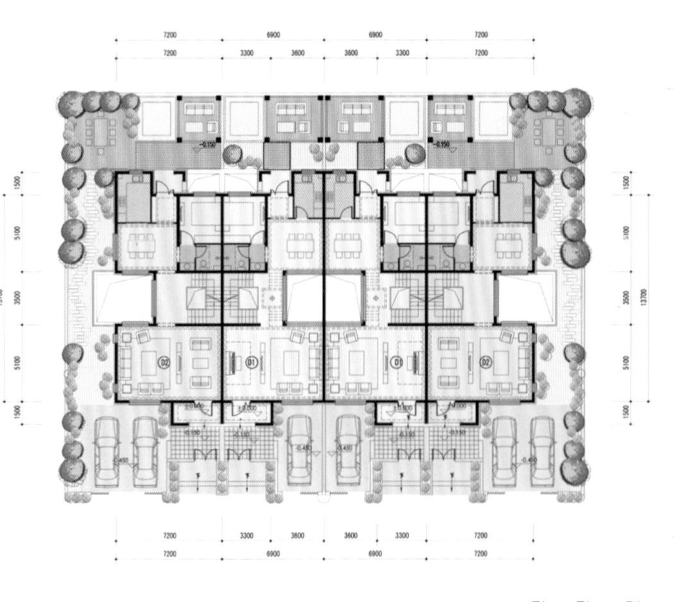

First Floor Plan
一层平面图

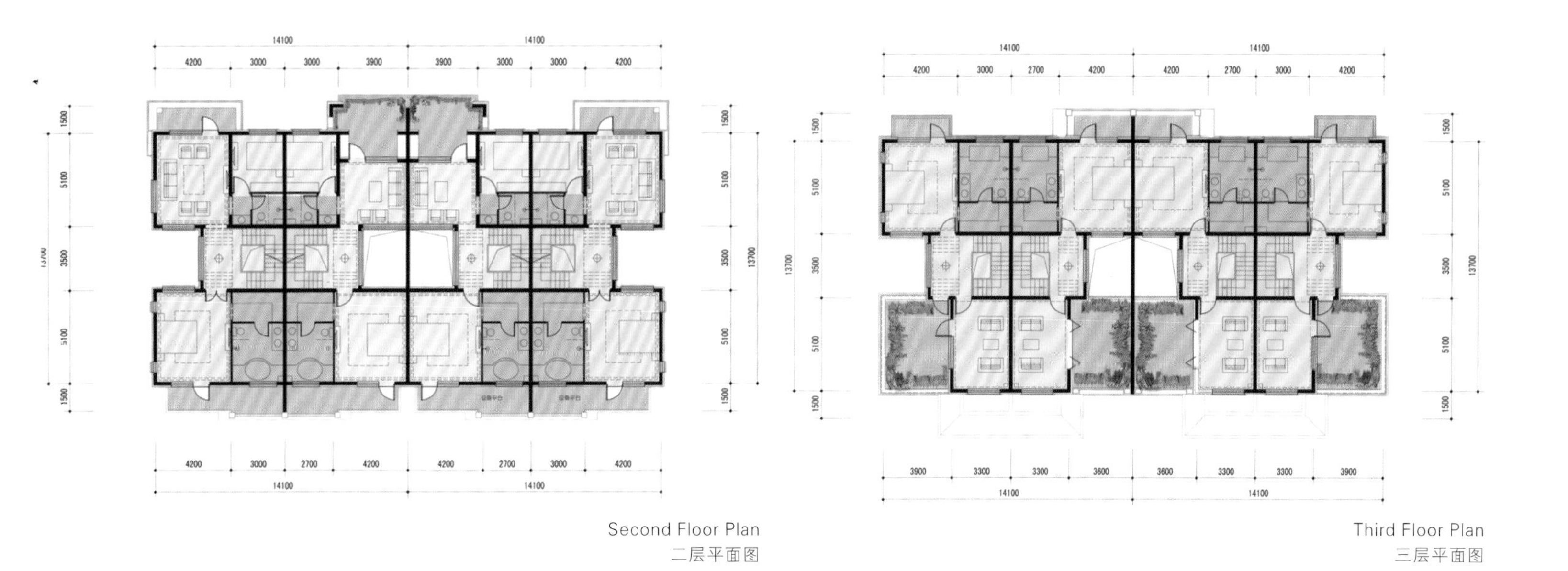

Second Floor Plan
二层平面图

Third Floor Plan
三层平面图

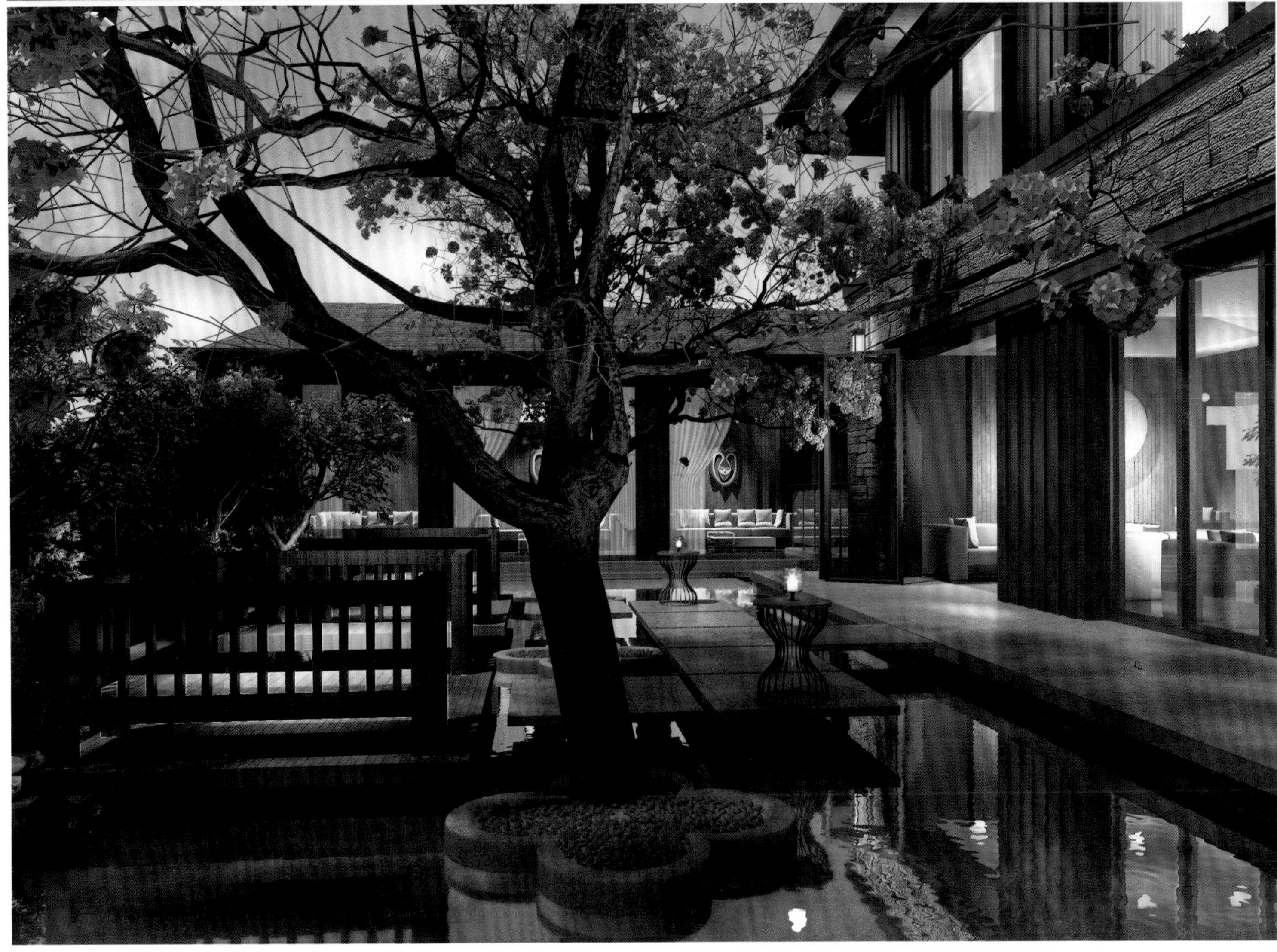

Landsea Meilizhou Garden

朗诗美丽洲

Location: Hangzhou, Zhejiang, China
Developer: Hangzhou Minglang Property Co., Ltd.
Architectural Design: Lacime Architectural Design
Land Area: 60,472 m^2
Floor Area: 48,377.6 m^2
Plot Ratio: 0.8
Green Coverage Ratio: 35%

项目地点：中国浙江省杭州市
开发商：杭州明朗置业有限公司
设计单位：上海日清建筑设计有限公司
占地面积：60 472 m^2
建筑面积：48 377.6 m^2
容积率：0.8
绿化覆盖率：35%

FEATURES 项目亮点

As a trendsetting product of Landsea G2 Collection, the project introduces up-to-the-minute world-class green building technologies and ideas as a way to build green and low-carbon buildings.

作为朗诗集团第二代产品的最高端系列，项目运用了大量代表世界前沿水准的绿色建筑技术和理念，代表了未来绿色低碳建筑的发展趋势。

KEYWORDS 关键词

Low-carbon Architecture
低碳建筑

Living Idea
生活理念

New-type Villa
新型亲地别院

Overview

As Landsea Real Estate's first pure low-density project in Hangzhou City and Landsea Group's updated sci-tech housing product, Landsea Meilizhou Garden is unparalleled in location, environment, planning, construction, and application of technology, and it is a real niche that harmonizes the environment, buildings, and human beings and creates a more natural living enviroment. In Landsea Meilizhou project there are only 185 villas. In a word, the Meilizhou project, a villa product integrating Mr. Wright's organic architectural theory with Tadao Ando's modernism, is a perfect marriage between green technology and nature, and it is devised only for the minority in the city who adore the nature.

项目概况

朗诗美丽洲是朗诗地产在杭州的首个低密度项目。其作为朗诗科技住宅的最新产品，在地段、环境、规划、建筑和科技运用等方面都有着无可比拟的优势，真正实现了“环境、建筑、人的和谐共生”，营造出“比自然更自然”的生活意境。朗诗美丽洲项目仅185席，集赖特的有机建筑理论与安藤忠雄的现代主义智慧于一身，是真正意义上的绿色科技与自然联姻的新型亲地别院，是城市里少有的亲近自然的项目。

Location Analysis

Nestled in the heart of Liangzhu, Landsea Meilizhou Garden has been fully integrated with Liangzhu Culture Village, and it leans against a primitive mountain range in the west and faces Meilizhou Park in the east, in addition to its complete supporting facilities such as Narada Five-star Resort and Anji Road Primary School etc. With the opening to traffic of Gudun Road Extension at the end of this year, Landsea Meilizhou Garden will have seamless connection with West Hangzhou City.

区位分析

朗诗美丽洲位于良渚圣地的核心区域，与良渚文化村连为一体。西依原生山脉，东望美丽洲公园，更有君澜五星度假酒店和安吉路小学等完美配套。随着古墩路延伸段的年底开通，与杭州城西实现无缝对接。

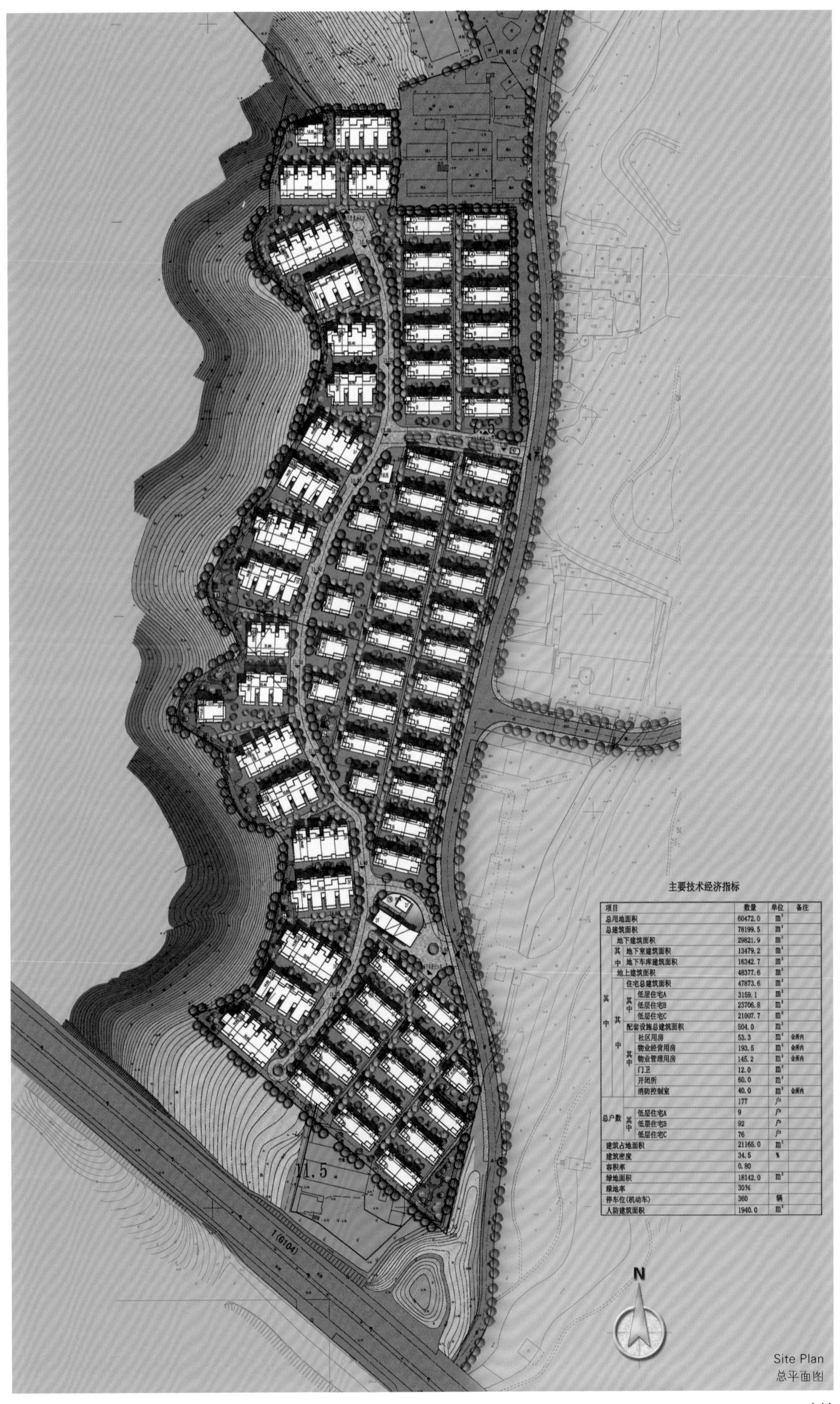

主要技术经济指标

项目	数量	单位	备注
总用地面积	60472.0	m²	
总建筑面积	78199.5	m²	
其中 地下建筑面积	29821.9	m²	
其中 地下室建筑面积	13479.2	m²	
其中 地下车库建筑面积	16342.7	m²	
地上建筑面积	48377.6	m²	
其中 住宅总建筑面积	47873.6	m²	
其中 低层住宅A	3159.1	m²	
低层住宅B	23706.8	m²	
低层住宅C	21007.7	m²	
配套设施总建筑面积	504.0	m²	
其中 社区用房	53.3	m²	会所内
物业经营用房	193.5	m²	会所内
物业管理用房	145.2	m²	会所内
门卫	12.0	m²	
开闭所	60.0	m²	
消防控制室	40.0	m²	会所内
总户数	177	户	
其中 低层住宅A	9	户	
低层住宅B	92	户	
低层住宅C	76	户	
建筑占地面积	21165.0	m²	
建筑密度	34.5	%	
容积率	0.80		
绿地面积	18142.0	m²	
绿地率	30%		
停车位(机动车)	360	辆	
人防建筑面积	1940.0	m²	

Site Plan
总平面图

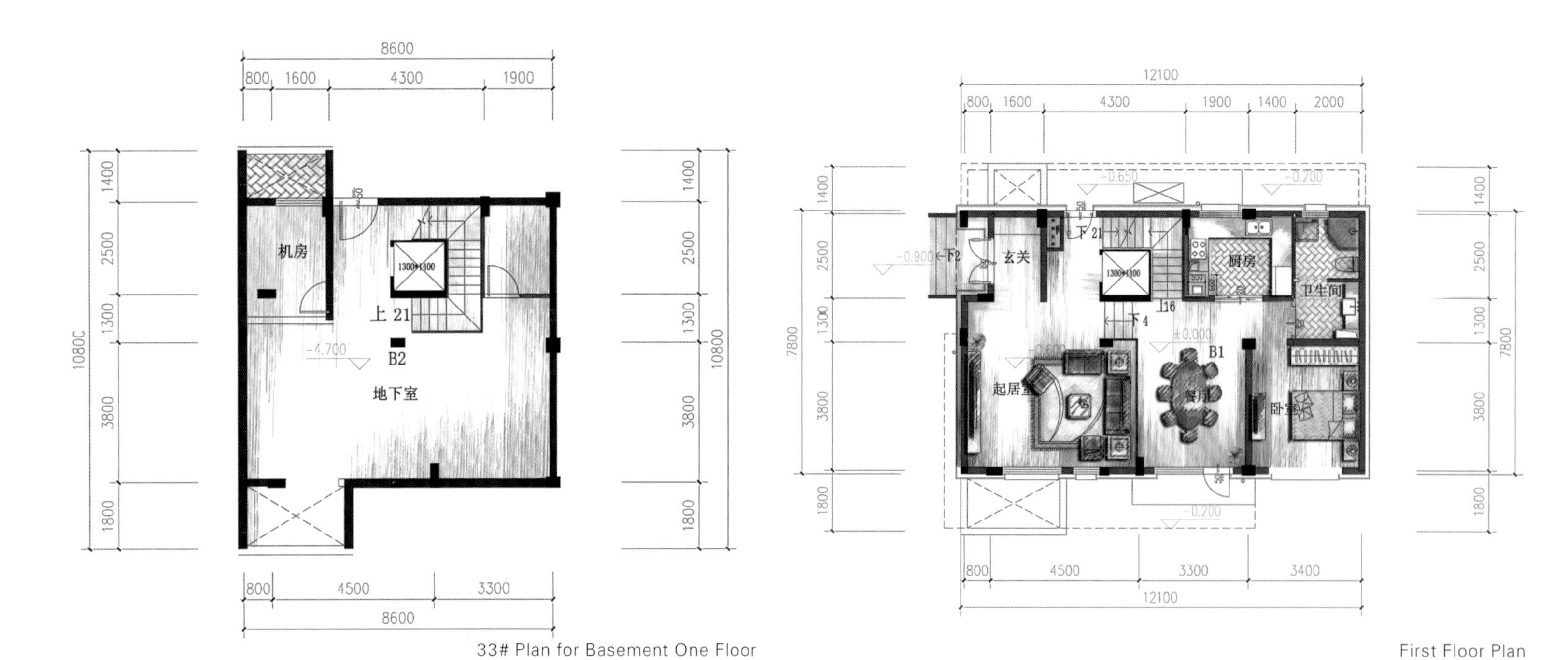

33# Plan for Basement One Floor
33# 楼地下一层平面图

First Floor Plan
一层平面图

Design Concept

The design uses advanced and reliable technologies, materials, structures and equipments to show Landsea's architectural idea of providing extensive customers with energy-saving, environmental-friendly, healthy, comfy, safe and low-carbon housing products and services. The reasonable distribution and application of varied materials embody the idea of sustainable development. And the planning and architectural design are based on and targeted at the idea of improving people's living environment. As a trendsetting product of Landsea G2 Collection, the project introduces up-to-the-minute world-class green building technologies and ideas as a way to build green and low-carbon buildings.

设计理念

设计采用先进且可靠的技术、材料、结构形式和设备，体现朗诗地产所倡导的"节能、环保、健康、舒适、安全"的建筑理念，打造低碳建筑、生活理念，合理分配和使用各项资源，全面体现可持续发展的思想，把提高人居环境质量作为规划设计、建筑设计的基本出发点和最终目的。作为朗诗集团第二代产品的最高端系列，项目运用了大量代表世界前沿水准的绿色建筑技术和理念，代表了未来绿色低碳建筑的发展趋势。

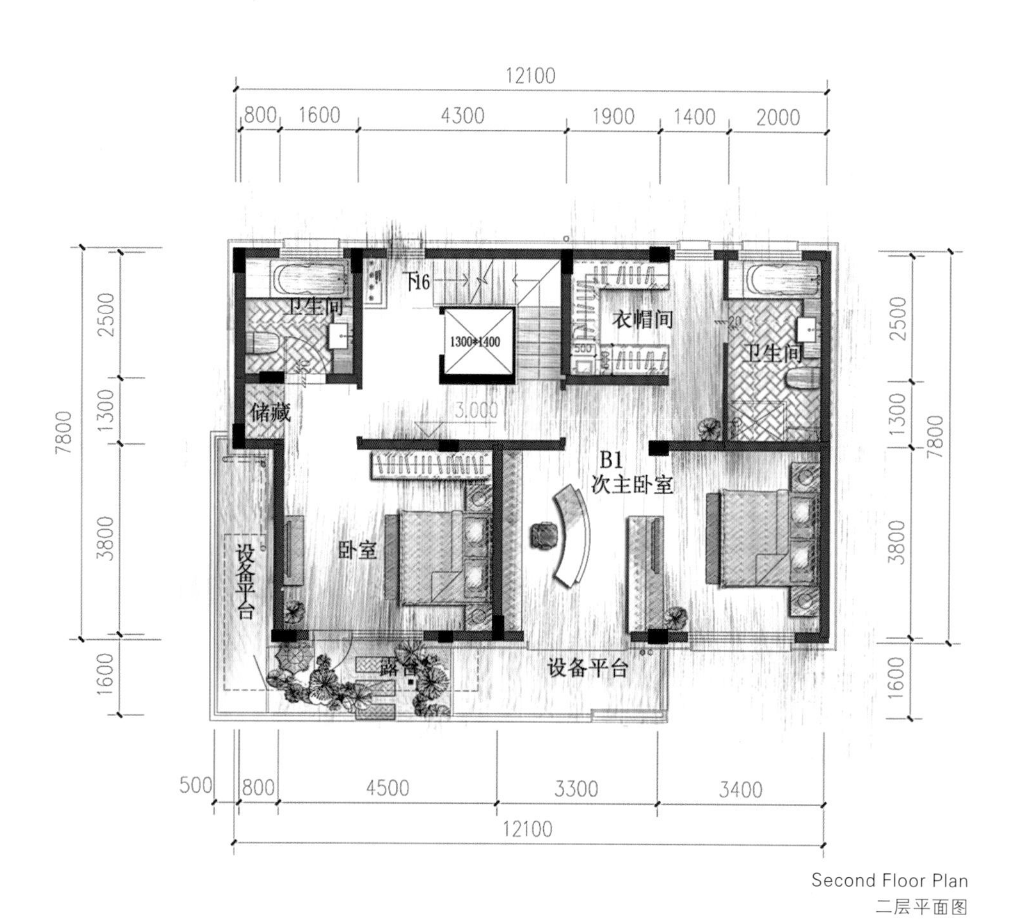

Second Floor Plan
二层平面图

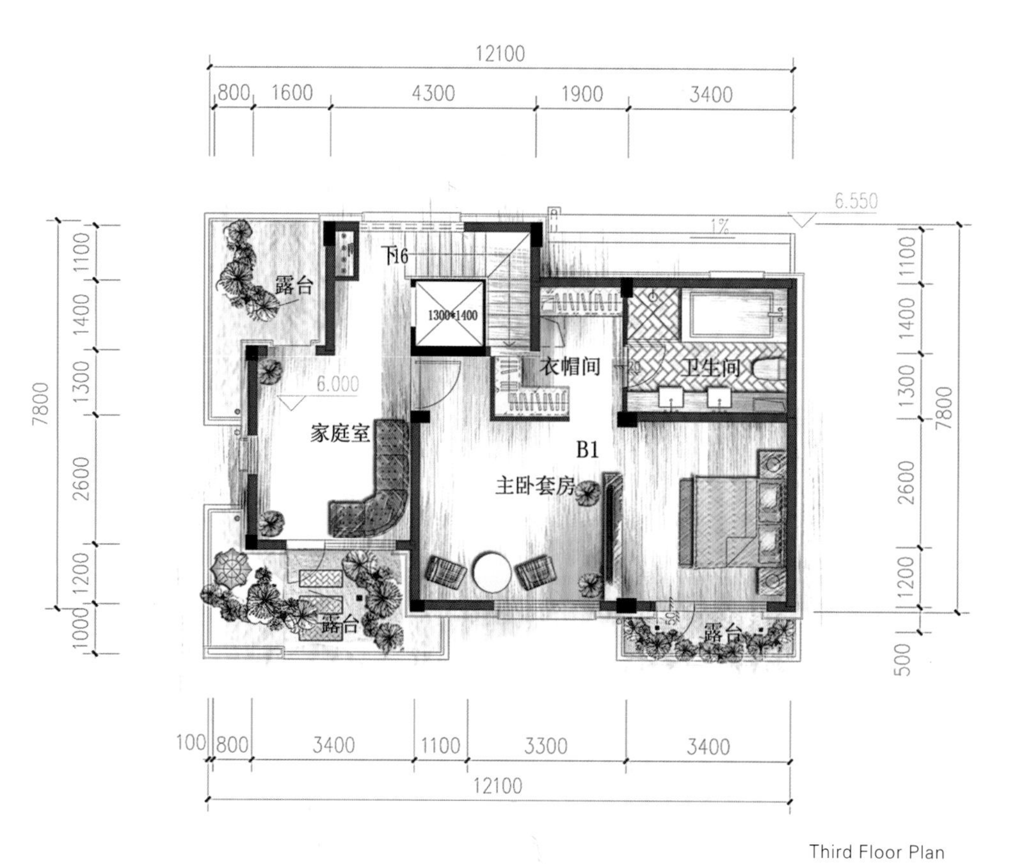

Third Floor Plan
三层平面图

CAREC Sanctuary

中航云玺大宅

Location: Kunming, Yunnan, China
Owner: CAREC (Kunming) Co., Ltd.
Architectural Design: Shenzhen Huahui Design Co., Ltd.
Land Area: 600,000 m^2
Floor Area: 600,000 m^2
Plot Ratio: 1.0

项目地点：中国云南省昆明市
业主：昆明中航房地产有限公司
设计单位：深圳华汇设计有限公司
占地面积：600 000 m^2
建筑面积：600 000 m^2
容积率：1.0

KEYWORDS 关键词

Traditional Space
传统空间

Prairie Style
草原风情

Refining Internally and Externally
内外兼修

FEATURES 项目亮点

Based on Wright's prairie style which is characterized by the oriental complex, the architectural style introduces more Chinese cultural elements to interpret traditional Chinese living ideas from space, architecture and details. It makes the project significant to our times.

建筑风格在赖特颇具东方情结的草原风格的基础上被赋予更多中国文化的元素，从空间、建筑、细部几个层面诠释中国传统居住理想，并赋予其一定的时代意义。

Overview

Located by the Dianchi Lake, CAREC Sanctuary is adjacent to the Wujiatang Wetland Park, enjoying a favorable natural environment.

项目概况

中航云玺大宅地处滇池之畔，紧临五甲塘湿地公园，自然环境优越。

Design Concept

Due to the height limitation of 12 m and the requirements of the owner for low-density villa quantity, the vast area still has limited space to bring into play. In the traditional Chinese living space, the residence is as important as the courtyard with the former representing the connotation and the latter representing the extension, which is also the traditional Chinese understanding to the world — refining internally and externally. In this project, the architects hope to combine this traditional Chinese idea on living with the modern idea of life. Therefore, the ground is designed to be used for walking and the parking lot to be placed underground. The sunken garden makes full use of the underground space and presents more landscape views to create a relaxing and pleasant atmosphere throughout the community.

设计理念

由于有12 m限高的要求，以及业主方对于低密度别墅总量的要求，虽然场地很大，但能发挥的空间是很有限的。在中国传统居住空间中，宅与院同等重要，一个是内涵，一个是外延，这也是中国传统对于世界的理解，讲究内外兼修。在这个项目里，设计师也希望把中国的传统居住理想与现代的生活理念相结合，为此，设计师首先把地面还给居民人行使用，所有停车在地下停车场解决，而下沉庭院则把地下室空间景观化、地面化，从而为社区塑造轻松怡人的环境奠定了基础。

Site Plan
总平面图

➤ Architectural Style

Based on Wright's prairie style which is characterized by the oriental complex, the architectural style introduces more Chinese cultural elements to interpret traditional Chinese living ideas from space, architecture and details. It makes the project significant to our times.

➤ 建筑风格

建筑风格在赖特颇具东方情结的草原风格基础上被赋予更多中国文化的元素，从空间、建筑、细部几个层面诠释中国传统居住理想，并赋予其一定的时代意义。

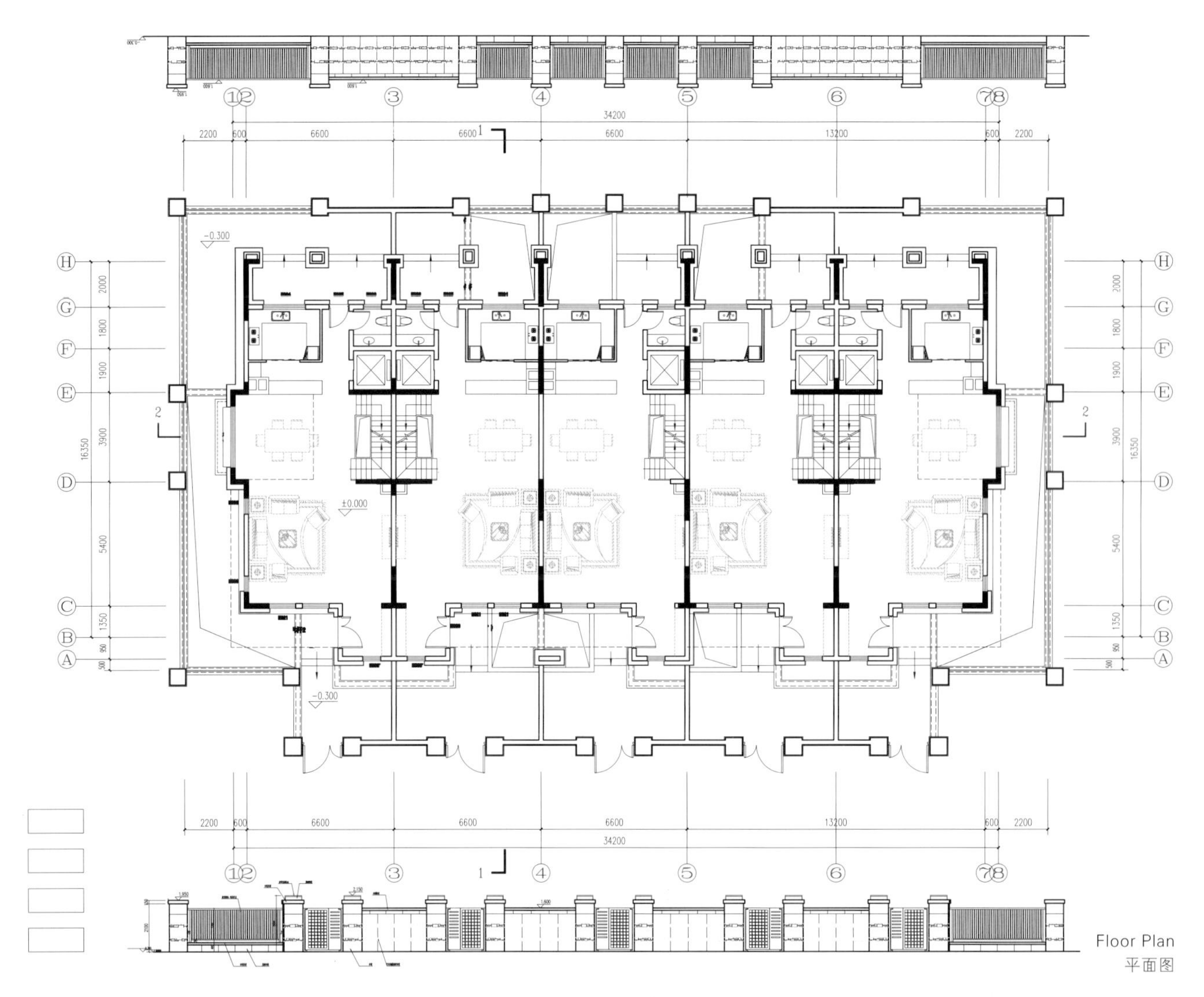

Floor Plan
平面图

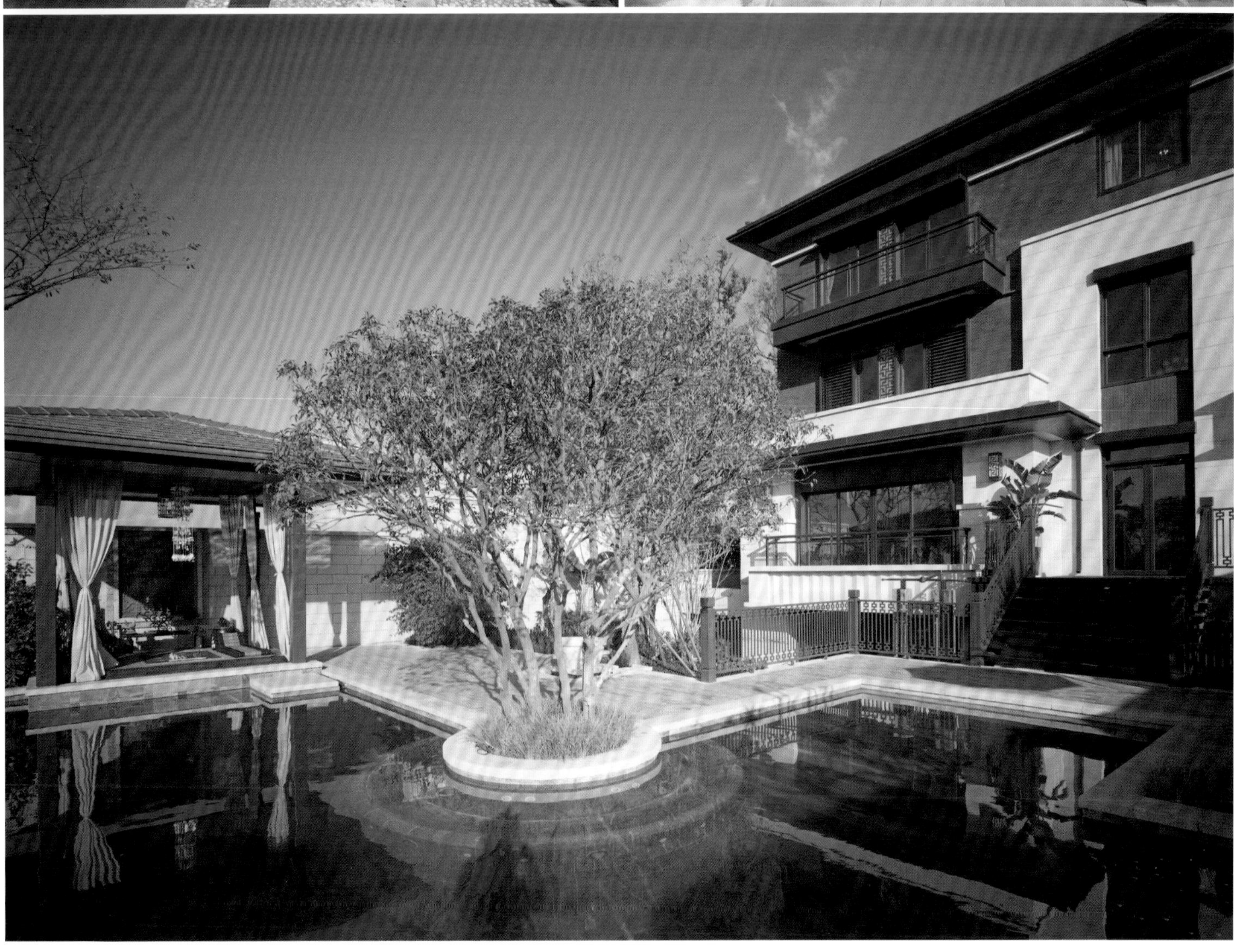

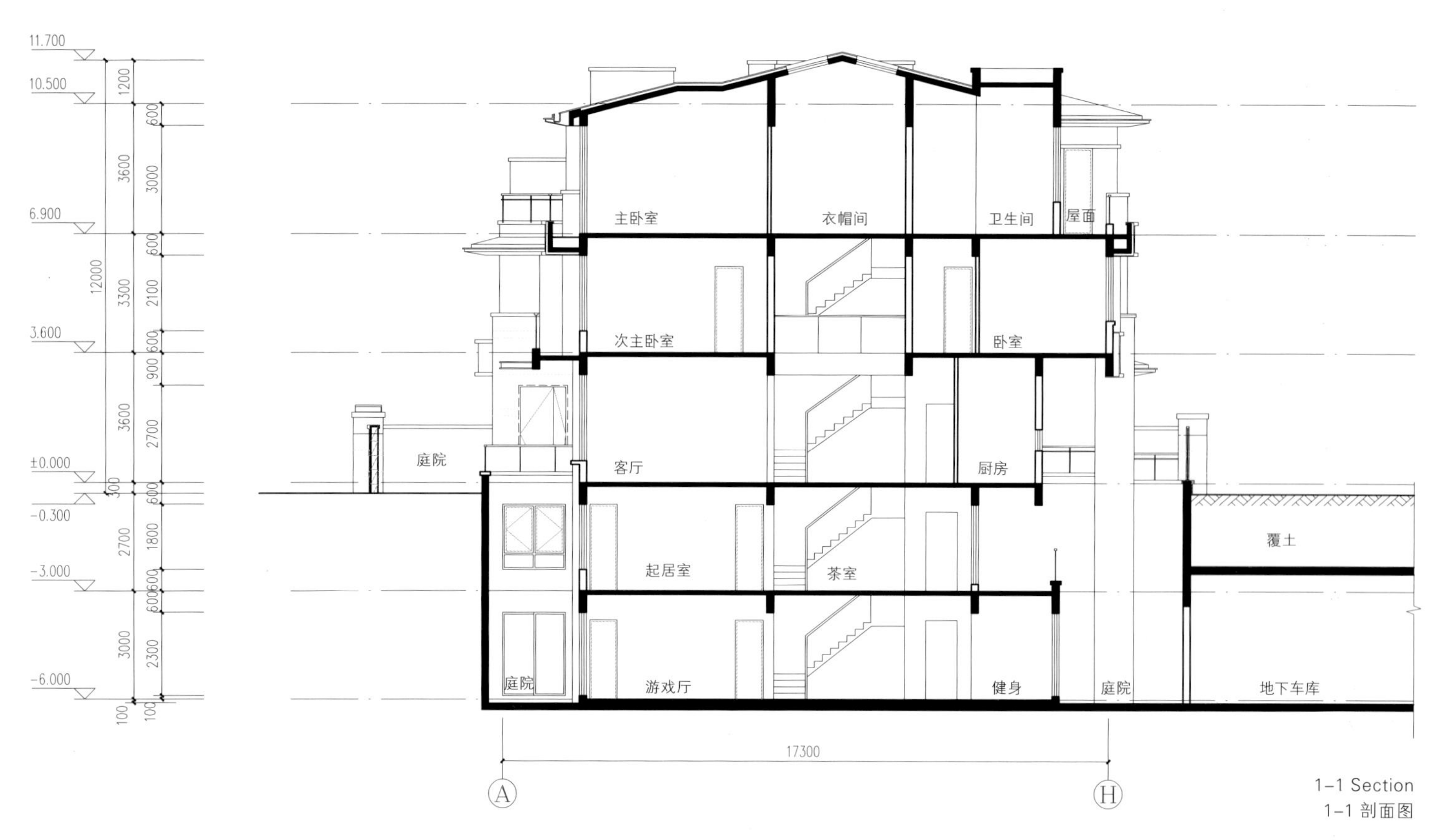

1-1 Section
1-1 剖面图

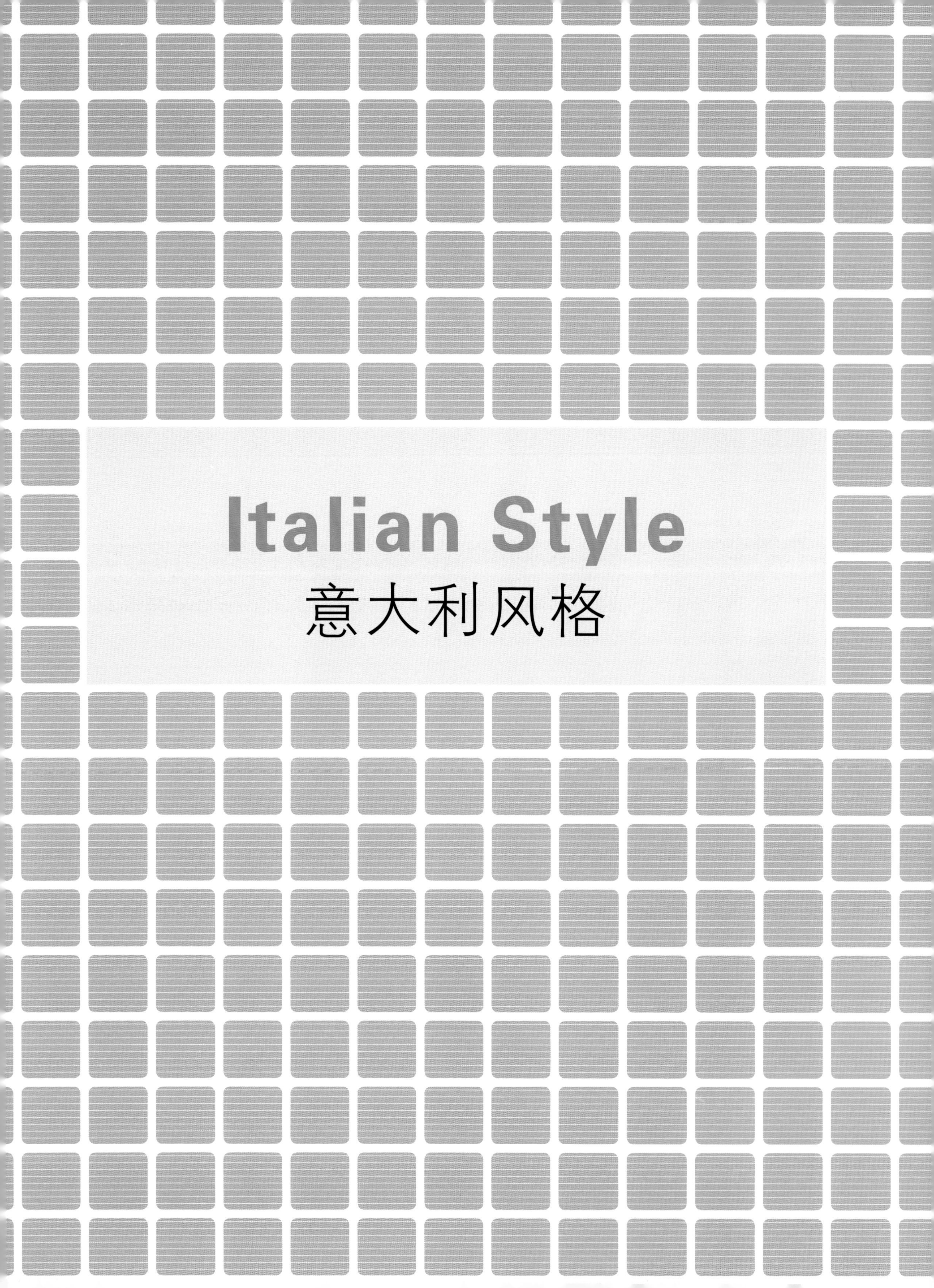

Italian Style
意大利风格

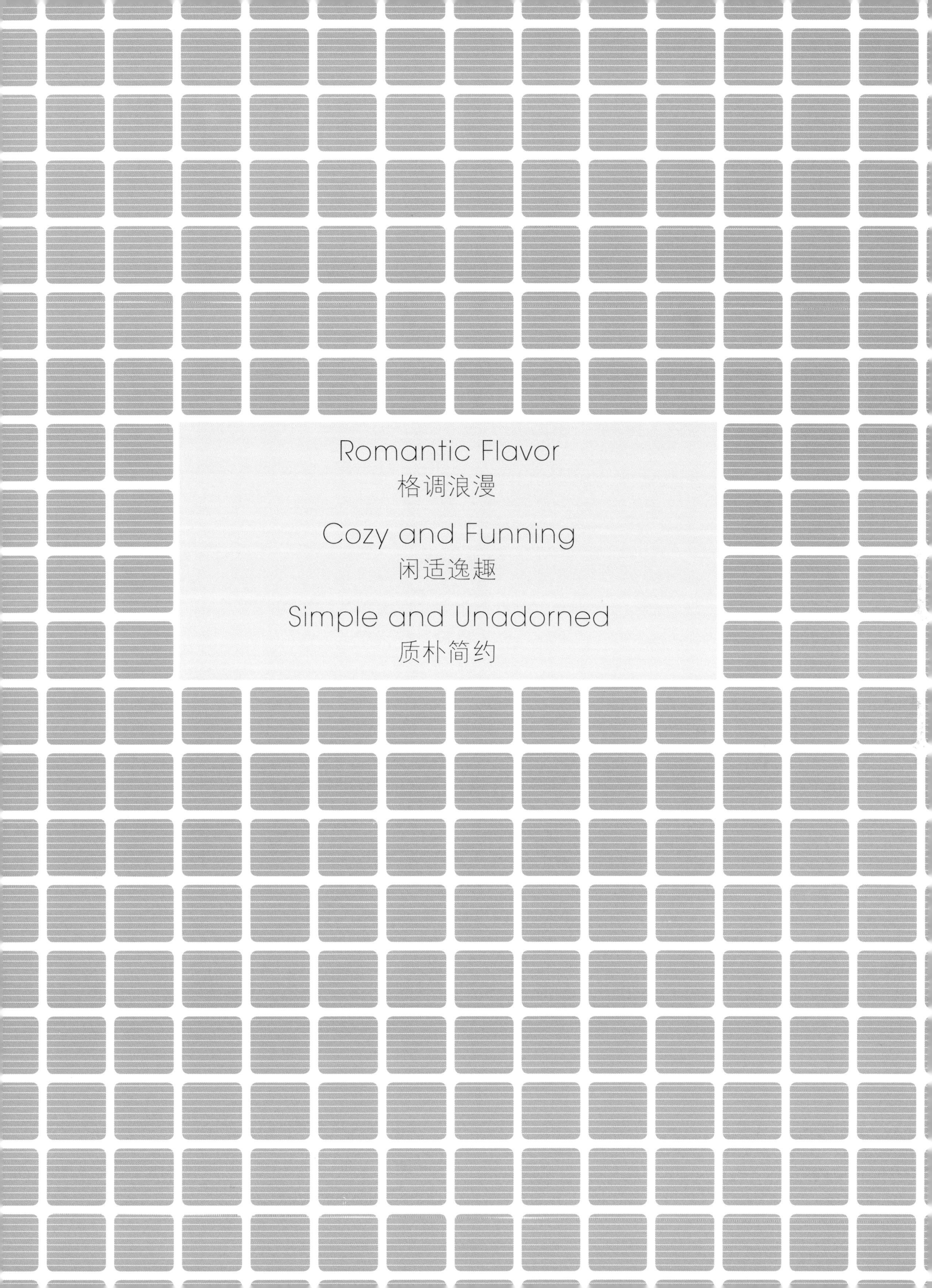

Romantic Flavor

格调浪漫

Cozy and Funning

闲适逸趣

Simple and Unadorned

质朴简约

Sansheng · Tuscany
三盛 · 托斯卡纳

Location: Fuzhou, Fujian, China
Developer: Sansheng Real Estate Group
Landscape Design:DDON Associates Planning and Design Co.,Ltd
Architectural Design:Sunlay Design
Land Area: 400,000 m^2
Floor Area: 299,997 m^2
Plot Ratio: 1.50
Green Coverage Ratio: 43%

项目地点：中国福建省福州市
开发商：三盛地产集团
景观设计：笛东联合（北京）规划设计顾问有限公司
设计单位：三磊设计
占地面积：400 000 m^2
建筑面积：299 997 m^2
容积率：1.50
绿化覆盖率：43%

KEYWORDS 关键词

Tuscany
托斯卡纳

Wonderful Scenery
极致风光

Native Community
原乡社区

FEATURES 项目亮点

Sansheng · Tuscany is a River · Mountain · Lake native community developed by Sansheng Real Estate Group. It creates three delicate natural landscapes, including "a 40,000 m^2 native lake, a 600 m long flower valley and 8 landscape corridors", which reproduce the wonderful scenery of Italian Tuscany.

三盛托斯卡纳是三盛地产集团倾力打造的江·山·湖原乡社区，并精心淬炼出"40 000 m^2 原生湖泊、600 m 花谷、8 条景观廊道"三重精致天然园景，再现了意大利托斯卡纳的极致风光。

Overview

Sansheng · Tuscany is a River · Mountain · Lake native community built by Sansheng Real Estate Group. It creates three delicate natural landscapes, including "a 40,000 m^2 native lake, a 600 m long flower valley and 8 landscape corridors", which reproduce the wonderful scenery of Italian Tuscany.

项目概况

三盛·托斯卡纳是三盛地产集团倾力打造的江·山·湖原乡社区，精心淬炼出"40 000 m^2原生湖泊、600 m花谷、8条景观廊道"三重精致天然园景，再现了意大利托斯卡纳的极致风光。

Design Concept

Sansheng · Tuscany integrates the city's prosperity and the natural beautiful scenery. It is one minute's drive away from the Gold Transaction Center and Qunsheng Jiangshancheng Complex, and five minutes' drive away from Cangshan Wanda Plaza and Outlets Premier Product Discount Store.

设计理念

三盛·托斯卡纳兼具城市繁华与自然美景。与黄金交易中心、群升江山城综合体仅1分钟车程，与仓山万达广场、奥特莱斯名品折扣城仅5分钟车程。

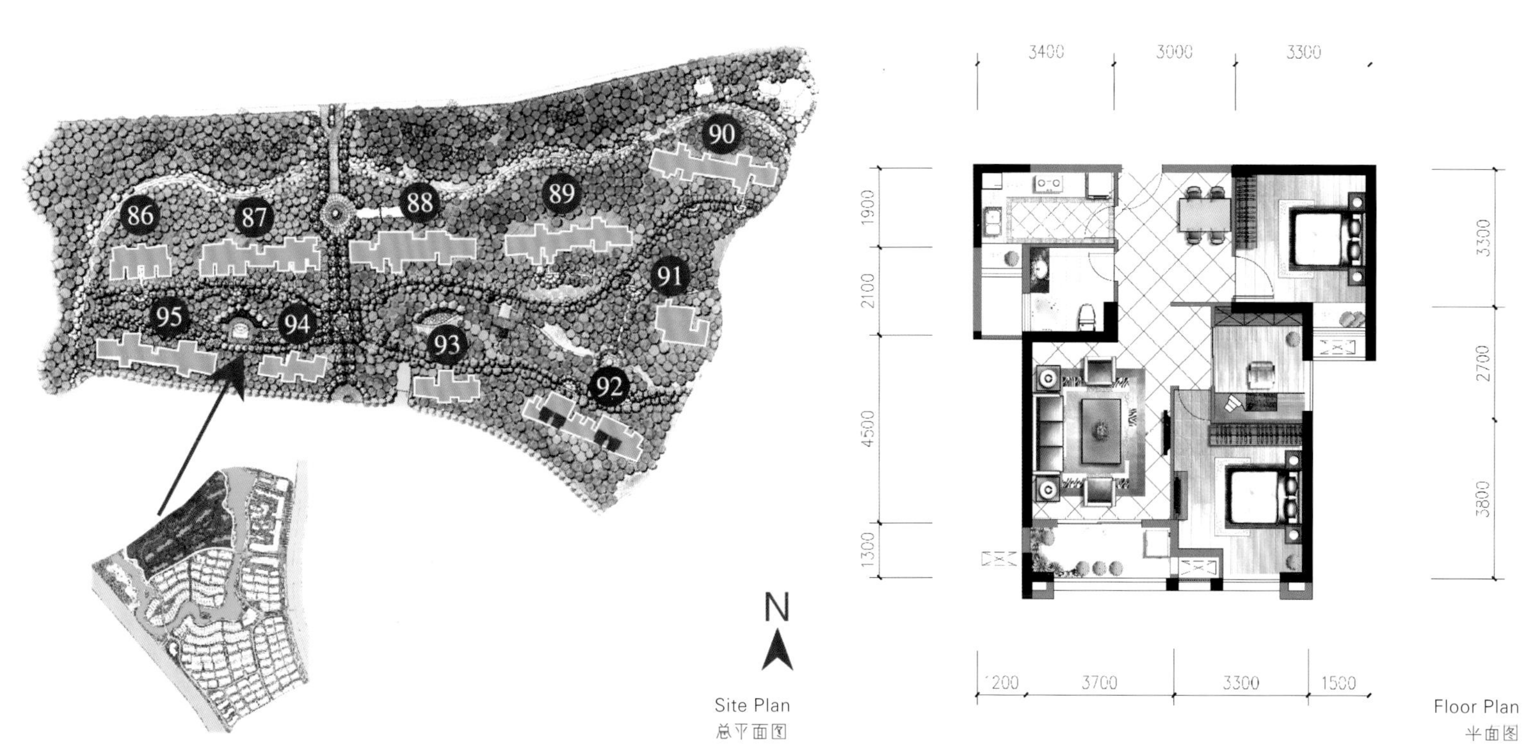

Site Plan
总平面图

Floor Plan
平面图

The lakefront villas are surrounded by an about 40,000 m^2 lake on three sides. The peripheral landscape of these villas is built with great pains; it takes water as the soul and flower as the pen, making the dual landscapes of "winding water and forest stream" and "romantic flower valley", providing every villa with its own unique landscape.

景观设计

三盛·托斯卡纳的一线湖岸独栋三面环绕60亩（约40 000 m^2）香湖。一线湖岸独栋在外围景观的营造上煞费苦心，它以水为魂，以花为笔，谱写出“曲水林溪”“浪漫花谷”的双重景致，为每栋别墅定制专属景观带，让每一户皆有独特的风景。

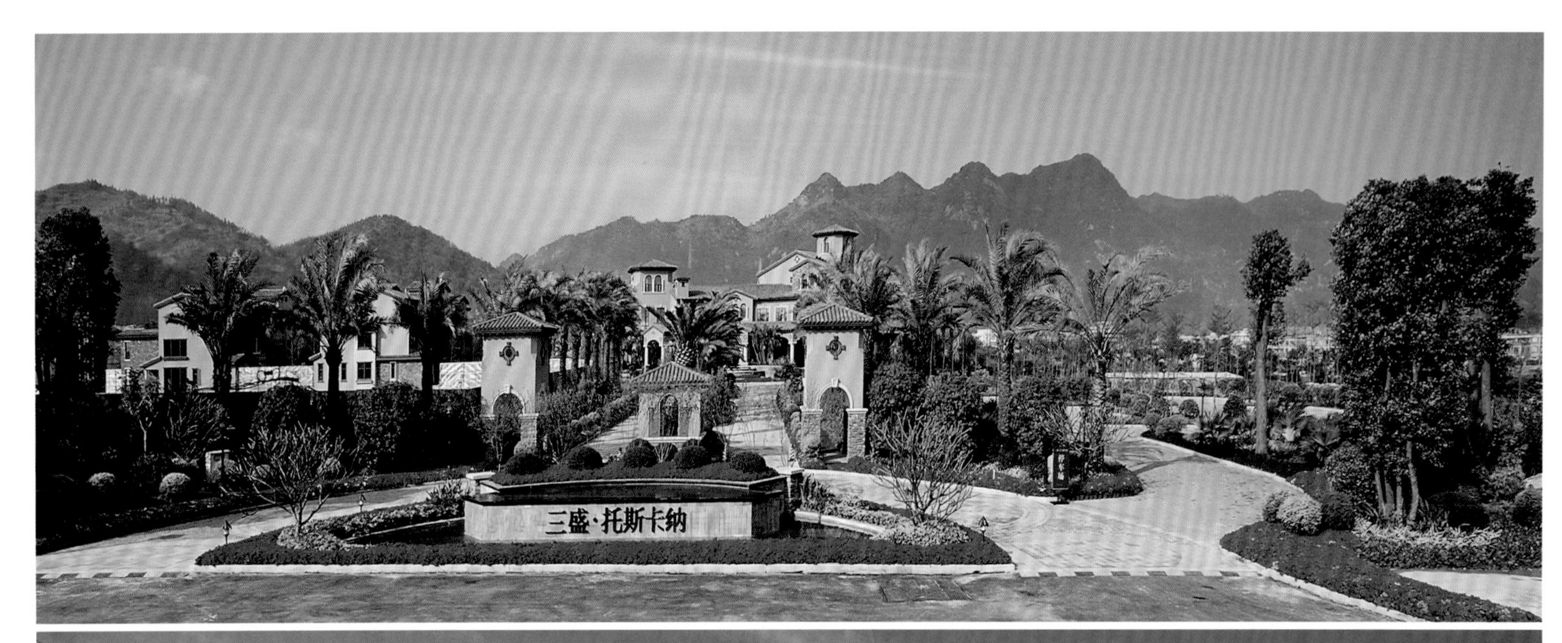

➤ Underground Palace

The luxury underground palace enjoyed by lakeshore villa has an area of 120 m^2, making you enjoy both the underground palace and the living experence on the ground floor, and it is also connected with the courtyard and the lake. What makes it more amazing is that the underground palace can be used as the study with hidden treasure, private wine cellar, audio & video room, etc. With extraordinary imagination, it overturns all the rules on villa construction.

➤ 地下宫殿

城市湖畔独栋尊享的"被抬高"的豪华地下宫殿，既拥有地下室约120 m^2的附赠价值，又享受首层的居住体验，它还与庭院、香湖相连。更令人惊叹的是，豪华地下宫殿可用作藏宝书房、私人酒窖、影音室等，注入的超凡想象力，颠覆了营造豪宅的全部规则。

➤ Additional Space

The additional space of lakeshore villa also includes a double parking space with an area of 85 m^2, which can accommodate two luxury private cars, and six oversized terraces and a double height living room. These additional spaces make the villa own a wide luxury space of 650 m^2 to 900 m^2.

➤ 附赠空间

城市湖畔独栋的附赠空间还包含约85 m^2的双车位，可停放两辆豪华私家车，以及6个超大花香露台和宫殿感挑空客厅等，这些附赠空间让城市湖畔独栋享受到650 m^2至900 m^2的奢阔空间。

Rongqiao City
融侨城

Location: Fuqing, Fujian, China
Developer: Rongqiao Real Estate Development Co., Ltd.
Architectural Design: Beijing Zhonghuajian Planning Design Institute Co., Ltd.
Landscape Design: EDAW Urban Design U+D

项目地点：中国福建省福清市
开发商：福清融侨房地产开发有限公司
设计单位：北京中华建规划设计研究院
景观设计：易道泛亚（香港）景观规划设计有限公司

FEATURES 项目亮点

International perspective and development concept are introduced to the project. It has created approx. 480,000 m^2 living area as the same as the foreign countries, making the project as the leader of quality life in Fuqing.

项目以国际化的视野与开发理念，将海外生活移植到福清，原汁打造了480 000 m^{2}的海外生活情景区，成为福清首屈一指的品质领先者。

KEYWORDS 关键词

Low Building Density
低密度

Unenclosed
合而不围

Grandeur and Spaciousness
恢宏大气

➤ Overview

The project is situated in Fuqing West District, which is the image window of politics, culture, commerce, finance, entertainment and residences of Fuqing. Having created approx. 480,000 m^2 living area as the same as the foreign countries, it has fully established the new construction idea of city plate and expressed the overseas quality of life exactly through the living atmosphere of Upper East Side, low building density, bionic landscape planning, gorgeous lake view, oversized volumes with 200 mu (approx. 133,333.33 m^2), as well as 480,000 m^2 urban complex.

➤ 项目概况

福清融侨城位于福清集政治、文化、商业、金融、娱乐和居住等为一体的形象窗口——福清西区。融侨地产打造了480 000 m^{2}的海外生活情景区。整个项目原汁原味的上东区国际居住氛围、恢宏大气的低密度建筑手笔、仿生自然化景观规划、静谧豪阔的唯一社区湖景、200亩（约133 333.33 m^2）超大体量、480 000 m^{2}的都市综合体开创了城市板块建设新理念，真实演绎了海外生活品质。

➤ Project Features

International perspective and development concept are introduced to the project, which has driven it as the leader of quality life in Fuqing. The three features of these building groups are being unenclosed, pretentious and distinguished. The residences with similar living atmosphere of villas, together with the modern skyscraper group composed of 18-layer high-rises and 33-layer high-rises, interpret the overseas lives perfectly.

➤ 项目特色

项目以国际化的视野与开发理念，将海外生活移植到福清，成为福清首屈一指的品质领先者。合而不围、一点点自负和卓尔不群是融侨城建筑组团的三大规划特色。类别墅的居住氛围与18层高层、33层超高层组成的现代摩天建筑集群，是对海外生活的完美诠释。

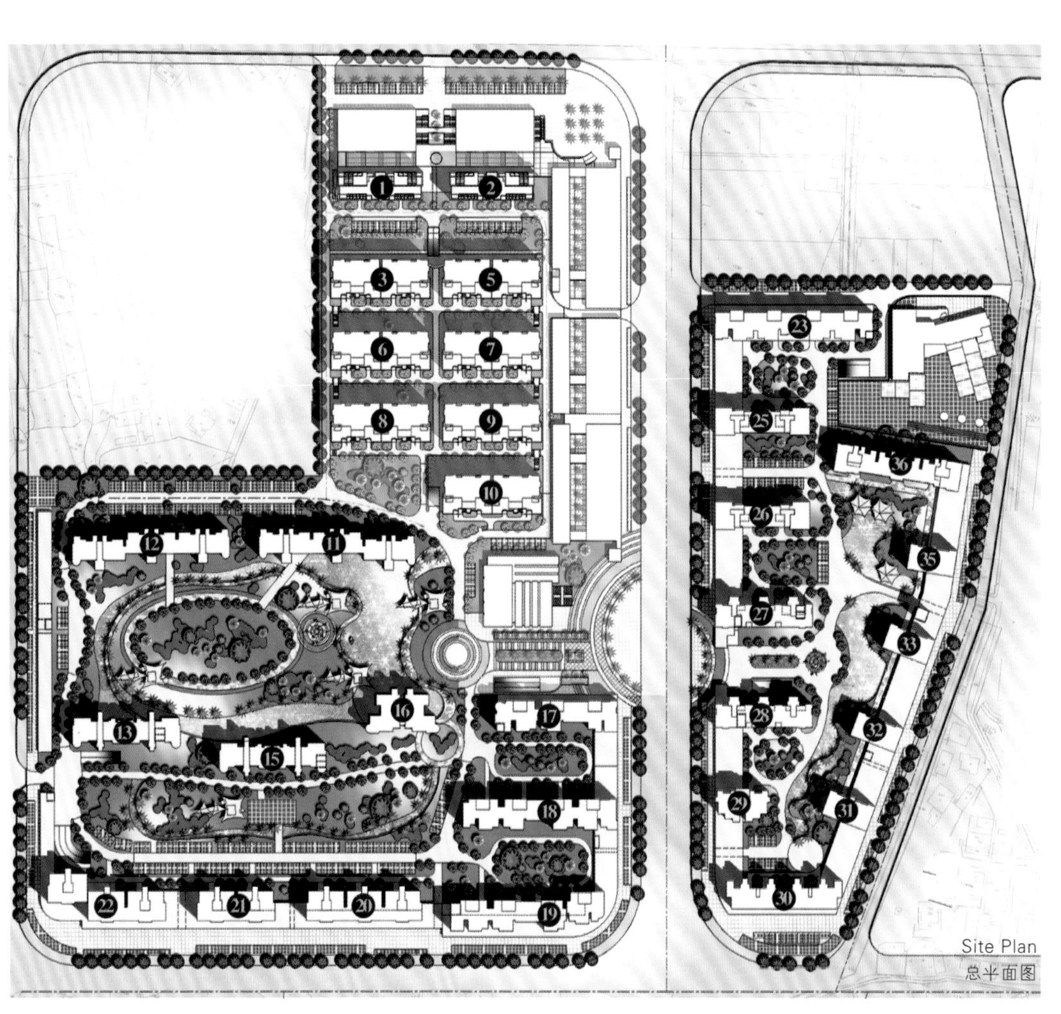

Site Plan
总平面图

➤ Overall Design

The phase Ⅰ is Lake County which is scene house not inferior to villa, establishing the template of quality life. The phase Ⅱ is Forest and Stream, with the lake all around and expressing the overseas life style. While the phase Ⅲ, Center Mansion, has absorbed the essences of the first 2 phases, to be designed more creatively with super-sized units and approx. 100 m height and distance between every two buildings which are quite impressive.

➤ 整体设计

一期湖郡，不亚于别墅的情景洋房，高调开启福清品质人居样板。二期林溪，全溪境泛会所式社区规划，完美诠释海外醇熟生活方式。融侨城三期中央公馆吸取一、二期精华，于产品设计上再次创新，纯粹大户，近百米高度，近百米栋距，惊世亮相。

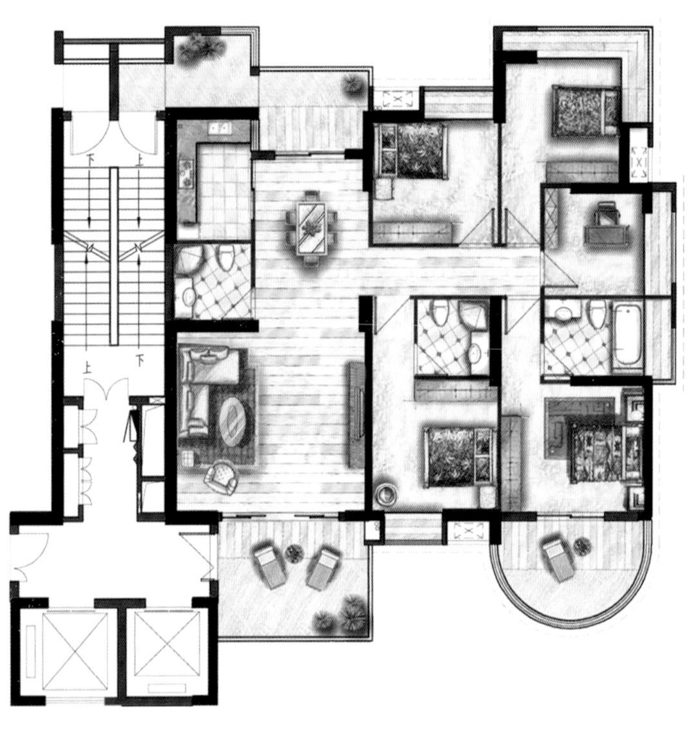

Floor Plan of Unit 01(Building 12)
12#01 单元平面图

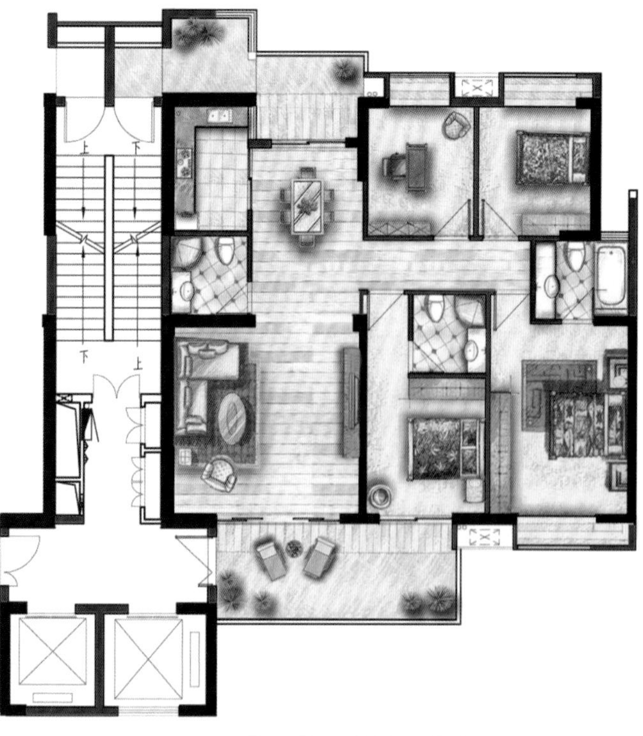

Floor Plan of Unit 05(Building 12)
12#05 单元平面图

Zone B of Luneng Jinan Residential Project, Maidao Village, Qingdao

青岛鲁能麦岛济南住区工程B区

Location: Qingdao, Shandong, China
Architectural Design: China Construction Engineering Design Group Corporation Limited
Gross Land Area: 152,300 m^2
Gross Floor Area: 356,100 m^2
Plot Ratio: 1.67
Green Coverage Ratio: 42.1%

项目地点：中国山东省青岛市
设计单位：中国中建设计集团有限公司
总占地面积：152 300 m^2
总建筑面积：356 100 m^2
容积率：1.67
绿化覆盖率：42.1%

KEYWORDS 关键词

Mountain-sea Views
透山透海

Seaside New Town
滨海新都市

Staggered Layout
错列布局

FEATURES 项目亮点

The planning for this residential area takes full advantage of the ideal location on the gold coast tourist line. The principle is clear on the concept of mountain-sea views, and ensures the sea views from the Hong Kong East Road and the wonderful views of Fushan National Forest Park from Donghai Road.

居住区规划设计最大限度地利用地处黄金海岸旅游线的地理特点，在理念上明确“透山透海”这一规划原则，实现既能够从香港东路看到大海，也可以从东海路看到浮山国家森林公园的城市观景。

Overview

Occupying a gross land area of 152,300 m^2, the project comprises four building groups. There are 3-storey villas, 6-storey garden apartments, 18- to 30-storey high-rises and 45-storey super high-rises (150 m high, with two additional floors underground), providing a gross floor area of 356,100 m^2.

项目概况

项目分为四个组团，总占地面积为152 300 m^2。项目总建筑面积为356 100 m^2，由3层别墅、6层花园洋房、18~30层高层及45层超高层组成；总建筑面积为356 100 m^2。另有地下2层。最高建筑高度为150 m。

Planning Design

The planning for this residential area takes full advantage of the ideal location on the gold coast tourist line, the principle is clear on the concept of mountain-sea views, and ensures the sea views from Hong Kong East Road, and the wonderful views of Fushan National Forest Park from Donghai Road. The buildings are arranged to be lower in the south and higher in the north to get the maximum views of the mountain and sea. At the same time, the height of the high-rises ranges from low to high, from west to east, to respond to the outline of Mount Fushan and form an undulating skyline. It has highlighted the space form in the seaside new town and introduced a healthy lifestyle among the mountain and sea.

规划设计

居住区的规划设计最大限度地利用地处黄金海岸旅游线的地理特点，在理念上明确“透山透海”这一规划原则，实现既能够从香港东路看到大海，也可以从东海路看到浮山国家森林公园的城市观景。规划布局采取南低北高的建筑群落，错列布局，争取最大朝向的观山、观海视线；同时高层建筑的高度随地势西低东高，呼应浮山的轮廓，形成高低起伏的天际线。项目体现出滨海新都市的空间形态，营造出山海间的健康生活。

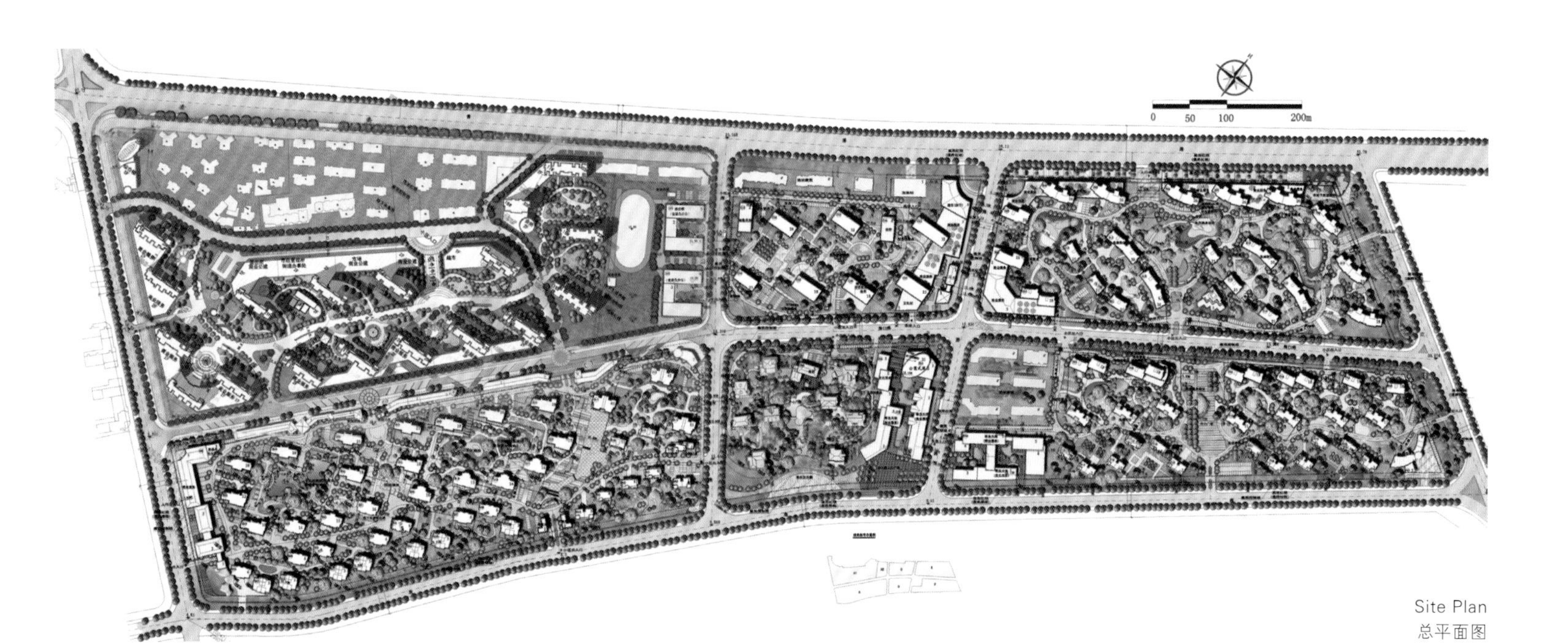

Site Plan
总平面图

➤ Plane Layout

The walkways are designed to be curved to blend into the surrounding undulating topography, green plants and gurgling streams. All these elements combine together to form a vertical landscape system which will properly separate the private spaces from the roads and the public green lands. At the same time, it will provide dynamic landscape effect for the community.

➤ 平面布局

在平面布局中人行道路尽量按照曲线形道路设计，两侧通过起伏地形、绿化植物、潺潺溪流的合理搭配形成景观形态和竖向高差的变化。由此起到隔离通行道路和绿地后侧的私密空间的作用，同时达到步移景异的景观效果。

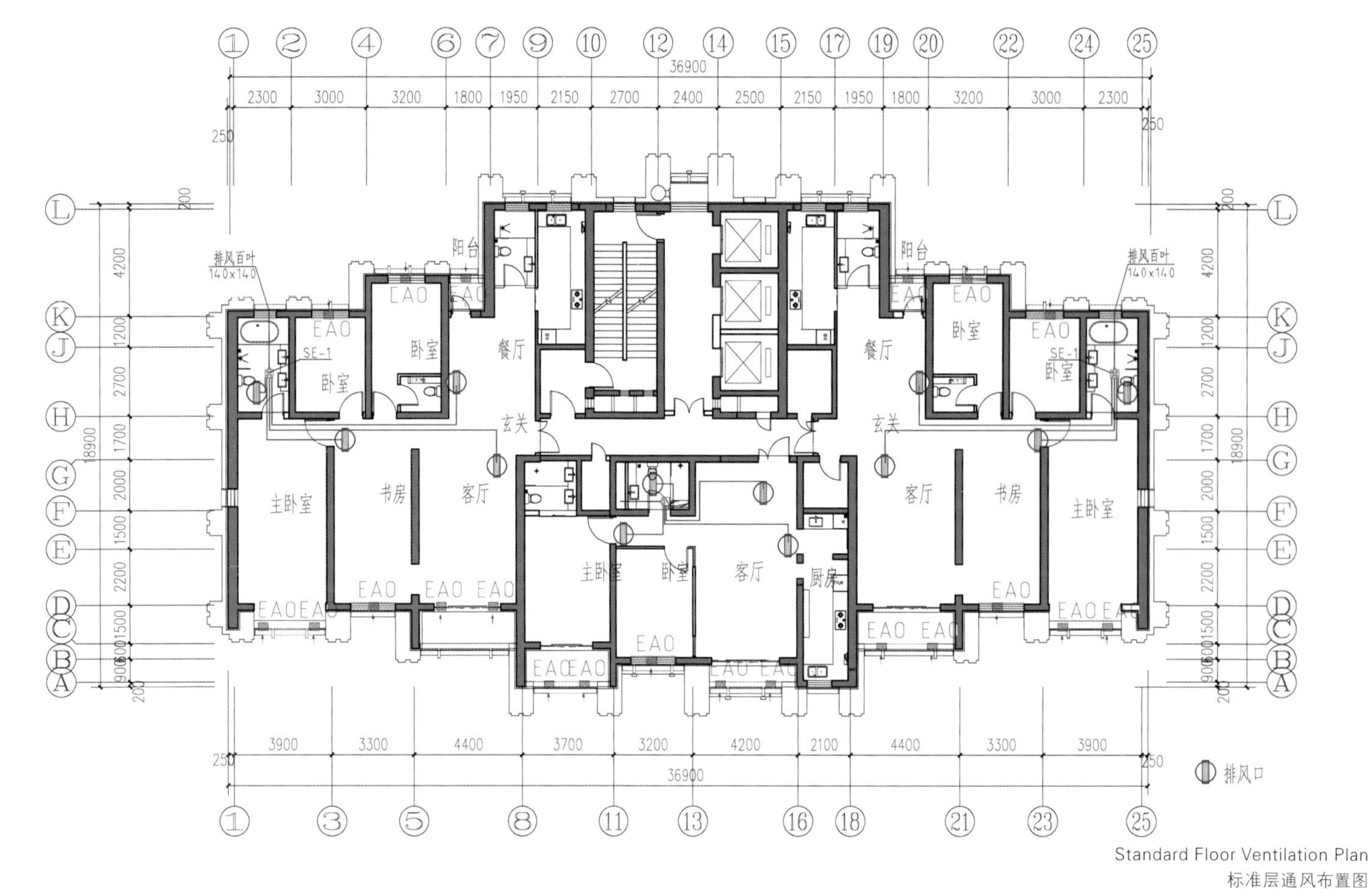

Standard Floor Ventilation Plan
标准层通风布置图

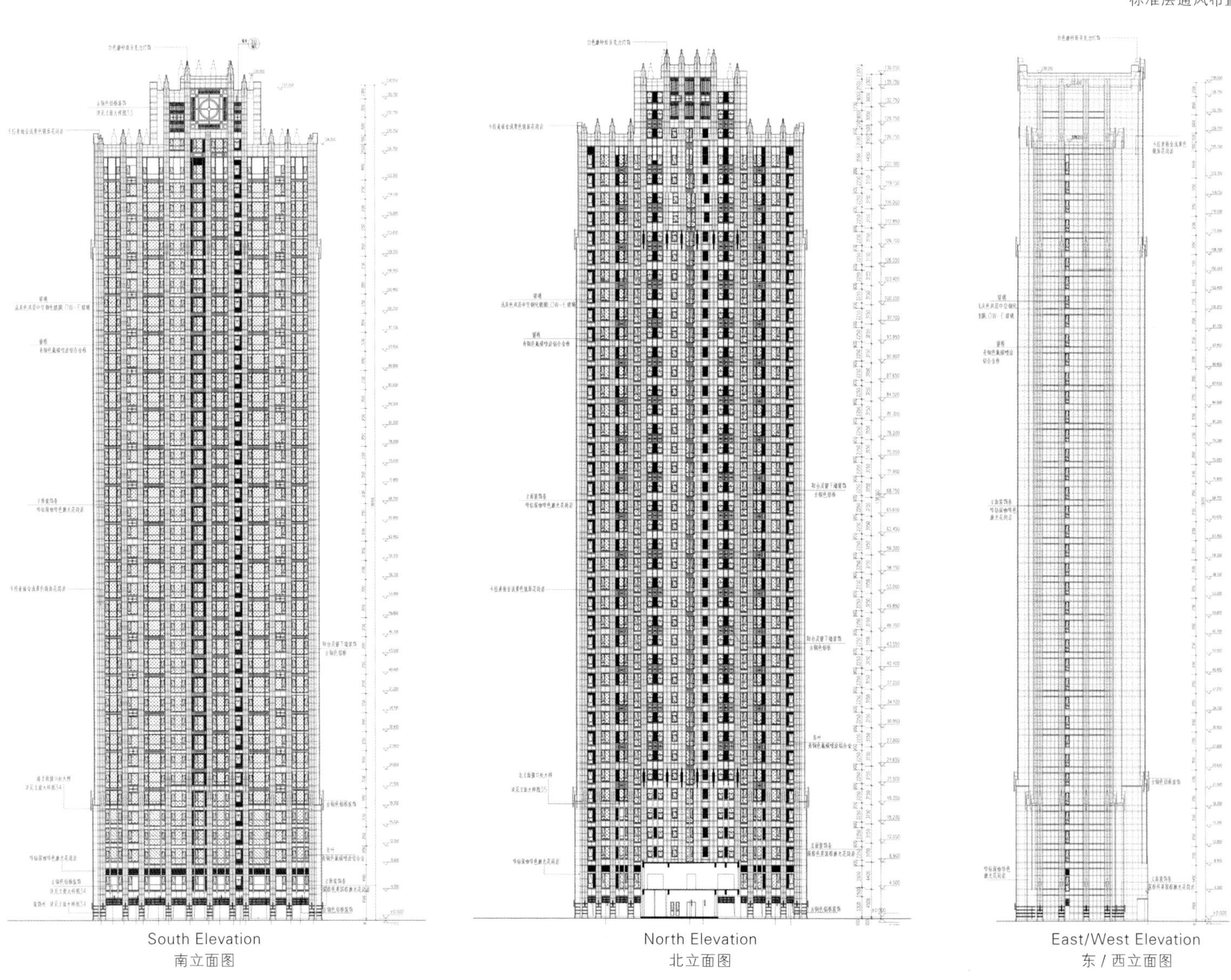

South Elevation
南立面图

North Elevation
北立面图

East/West Elevation
东 / 西立面图

French Style

法式风格

Vibrant and Unrestrained
活力奔放
Surprising and Elegant
惊喜别致
Magnificent
气势恢宏

Jiaxing Eastern Provence

嘉兴东方普罗旺斯

Location: Jiaxing, Zhejiang, China
Developer: Zhejiang Orient Blue Ocean Land Co., Ltd.
Landscape Design: Botao Landscape(Australia)
Total Land Area: 194,387 m^2
Total Floor Area: 440,302 m^2
Plot Ratio: 1.807
Green Coverage Ratio : 47.36%

项目地点：中国浙江省嘉兴市
开发商：浙江东方蓝海置地有限公司
景观设计：澳大利亚•柏涛景观
总占地面积：194 387 m^2
总建筑面积：440 302 m^2
容积率：1.807
绿地覆盖率：47.36%

FEATURES 项目亮点

All the low-rise residential projects adopt the L-shaped design and set up sinking courtyard and daylighting patio, thus the L-shaped plane can provide good daylighting and ventilation for underground.

项目所有低层住宅采用“L”型设计，并设下沉庭院及采光天井，“L”形平面使得地下室拥有良好的采光通风。

KEYWORDS 关键词

French Style
法式风格

L-Shaped Design
“L”型设计

Two-floor Terrace
双层高露台

Overview

The project is located in Dongzha Plot of South Lake New District, 3.8 km far away from the southeast of South Lake. The east of plot exists the extension part of Middle Ring East Road of this plan, south to the Shanghai-Hangzhou Expressway and high-speed railway's transportation junction. As the city's restriction to north and development to south, this project just locates on the main direction of city's development, taking obvious advantage on regional traffic.

项目概况

项目地处南湖新区东栅地块，位于南湖东南约3.8 km处。地块东侧为规划中的中环东路延长段，向南可与沪杭高速公路及高速铁路交通枢纽连接，随着城市的北控南进，本项目所处区域为城市发展的主要方向，区域交通优势非常明显。

Project Plan

The project design tries to make use of specific area and surrounding environment, inspired by canal tributary and themed as water, to create dynamic environment and harmonious community for the city. Through the study of the surrounding status and plan, the site is divided into two areas according to landscape resources' advantages and disadvantages. Southeastern area is low-rise residence with favorable exterior landscape, while northwestern area is high-rise residence, nearing industrial area, with inferior landscape resources. Roads and water systems are used to divide public construction area, high-rise and low-rise residential areas, which are independent with each other but organically linked by the entrance, landscape axis of high-rise area and water systems to form a very clear plan structure.

项目规划

本方案设计力求利用项目特定的区域及周边环境，以运河支流为灵感，以水为主题，为小区创造一个灵动的环境，为城市创造一个和谐的社区。通过对项目周边现状及规划的分析，将用地按景观资源优缺分为两个区域，东南片区为低层住宅，拥有优越的外部景观环境，西北片区为高层住宅，靠近工业区，景观资源相对较弱。利用道路和水系将小区划分为公建区、高层住宅区和低层住宅区，各区相对独立，通过入口和高层区的景观轴线以及水体将各区有机地联系起来，形成非常清晰的规划结构。

Site Plan
总平面图

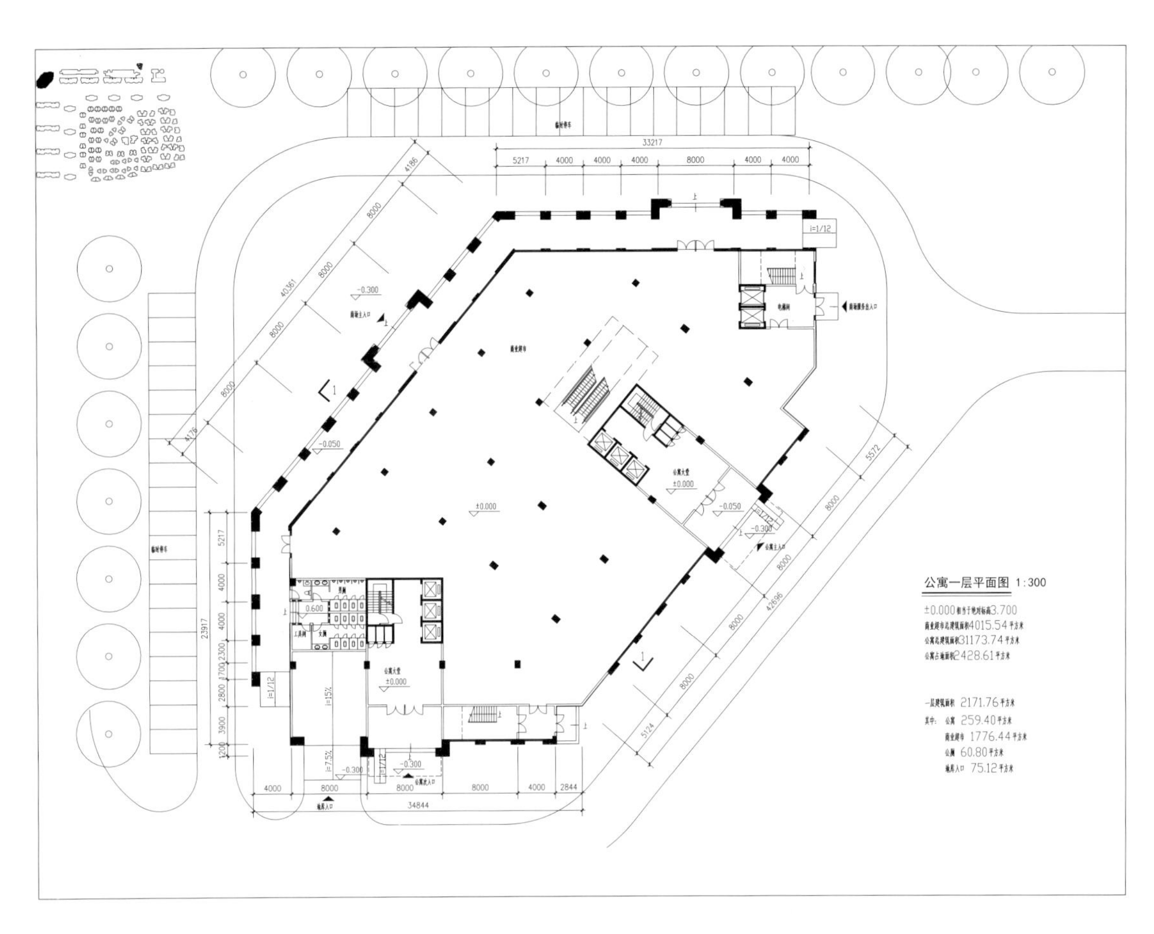

➤ Architectural Style

Based on French architectural style, the project's architectural form mainly reflects the ancient and exquisite European architecture. The pursuit of tranquil and comfortable living atmosphere keeps people far away from the urban hustle and bustle, while enjoying sunny, fresh and leisure life is the main life concept the designers trying to achieve. Low-rise residences and public buildings adopt quaint French style, showing elegance and dignity. High-rise residences adopt modern classical style, using stone, glass, steel and other modern materials and integrating classical elements to be classic and fashion.

➤ 建筑风格

项目建筑形式以法式建筑风格为基调，着重体现欧式建筑的古朴精致。追求宁静、舒适的居住气氛，让人们远离繁忙都市的喧嚣，尽情享受阳光、空气和悠闲的生活，是设计单位在设计中想努力带来的最主要的生活理念。低层住宅及公建采用古朴的法式风格，典雅高贵。高层住宅采用现代经典风格，利用石材、玻璃、钢等现代材料，在现代风格基础上融入古典元素，经典而时尚。

➤ Residential Design

High-rise residences consist of 8 buildings of 32-level unit-type slab floors in the form of two elevators with four or three households in each level and 8 buildings of 29-level duplex single house. House type's plane is design into flat plate floor, which compared with T-shaped plane enables each household to enjoy more sufficient sunlight since its form has no shield for sunshine. 4 buildings of unit-type high-rise slab floors in the form of two elevators with three households in each level are located in the north of the plot and 4 buildings of unit-type high-rise slab floors in the form of two elevators with four households in each level are in the west. All unit-type high-rise residences' house types have two living rooms except the duplex house in the top. Each household holds two-floor terraces in the north which greatly resolve the ventilation problem of house types in the middle of the buildings.

All the low-rise residential projects adopt the L-shaped design and set up sinking courtyard and daylighting patio, thus the L-shaped plane can provide good daylighting and ventilation for underground.

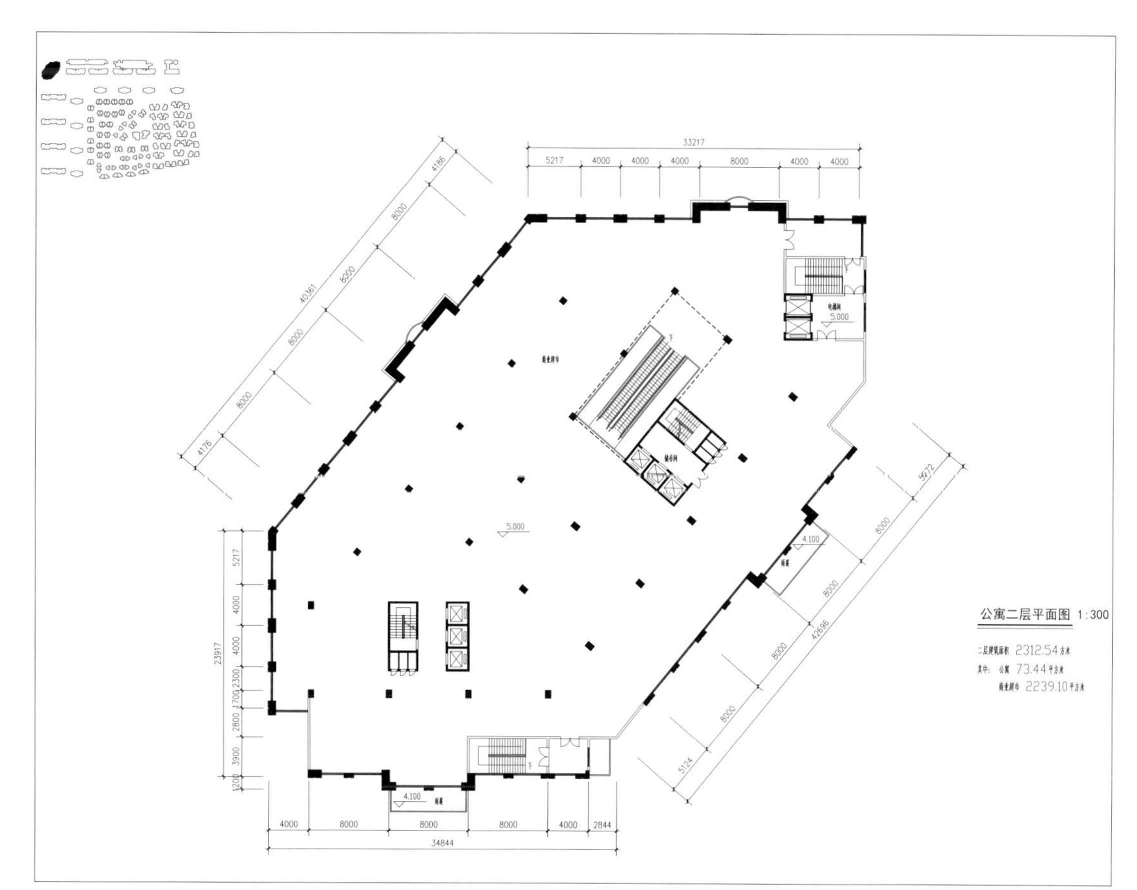

高层住宅由八栋32层的两梯四户和两梯三户单元式板楼以及八栋29层的复式独栋组成。户型平面采用平滑的板楼设计，相比“T”形平面，板楼的形式由于没有日照自遮挡，能使每户享受更充足的阳光。四栋两梯三户单元式高层板楼位于地块的北侧，四栋两梯四户高层位于地块西侧。除顶层复式外所有单元式高层住宅户型均为两房，每户拥有双层高大露台，中间户型通过北面双层高露台很好地解决户型通风问题。

项目所有低层住宅采用“L”形设计，并设下沉庭院及采光天井，“L”形平面使得地下室拥有良好的采光通风。

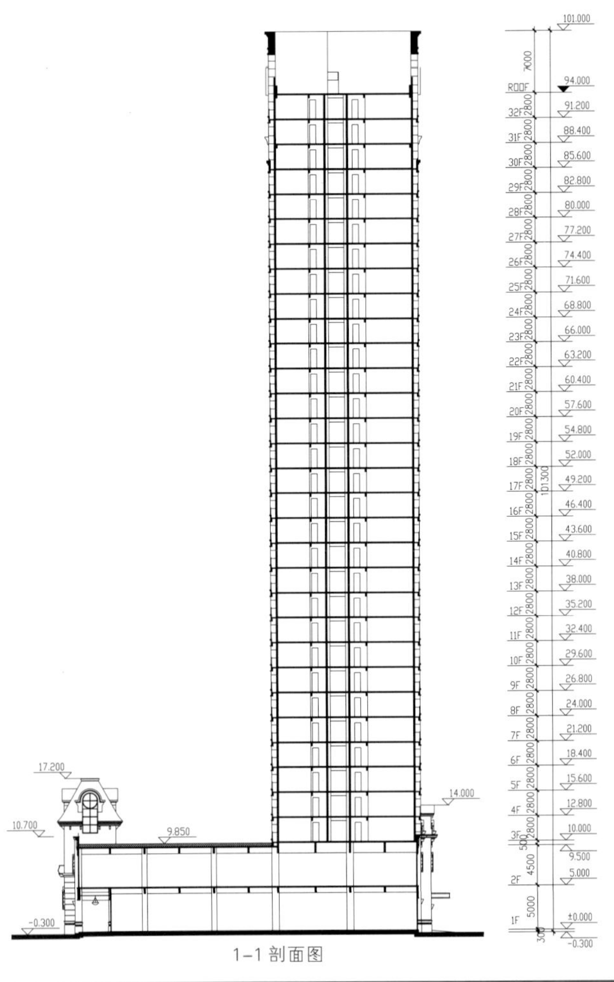

1-1 剖面图

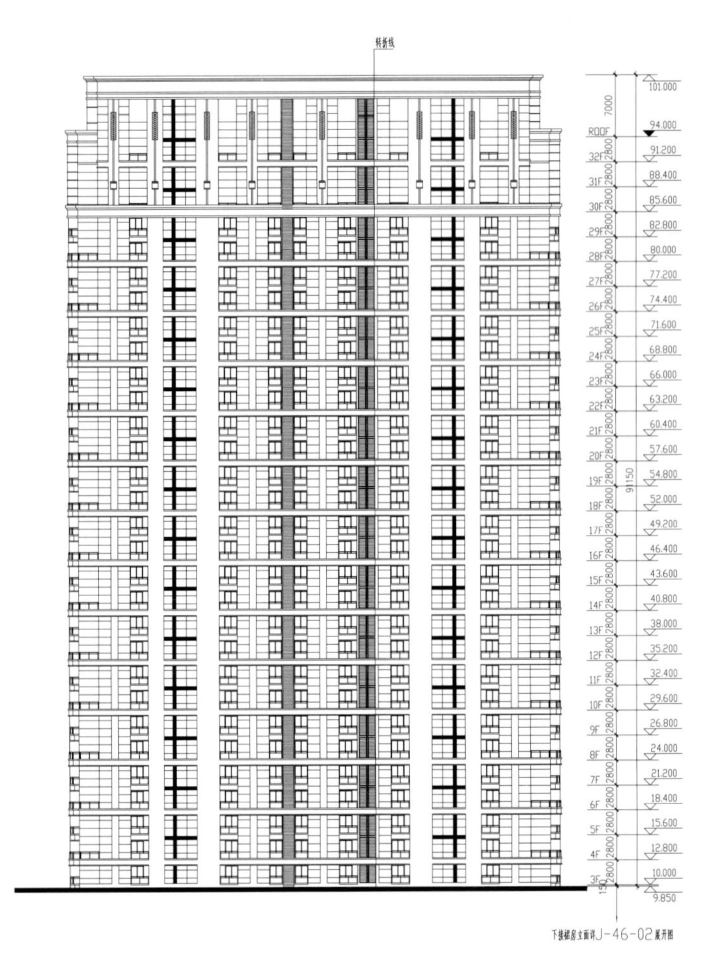

单身公寓塔楼北立面展开图

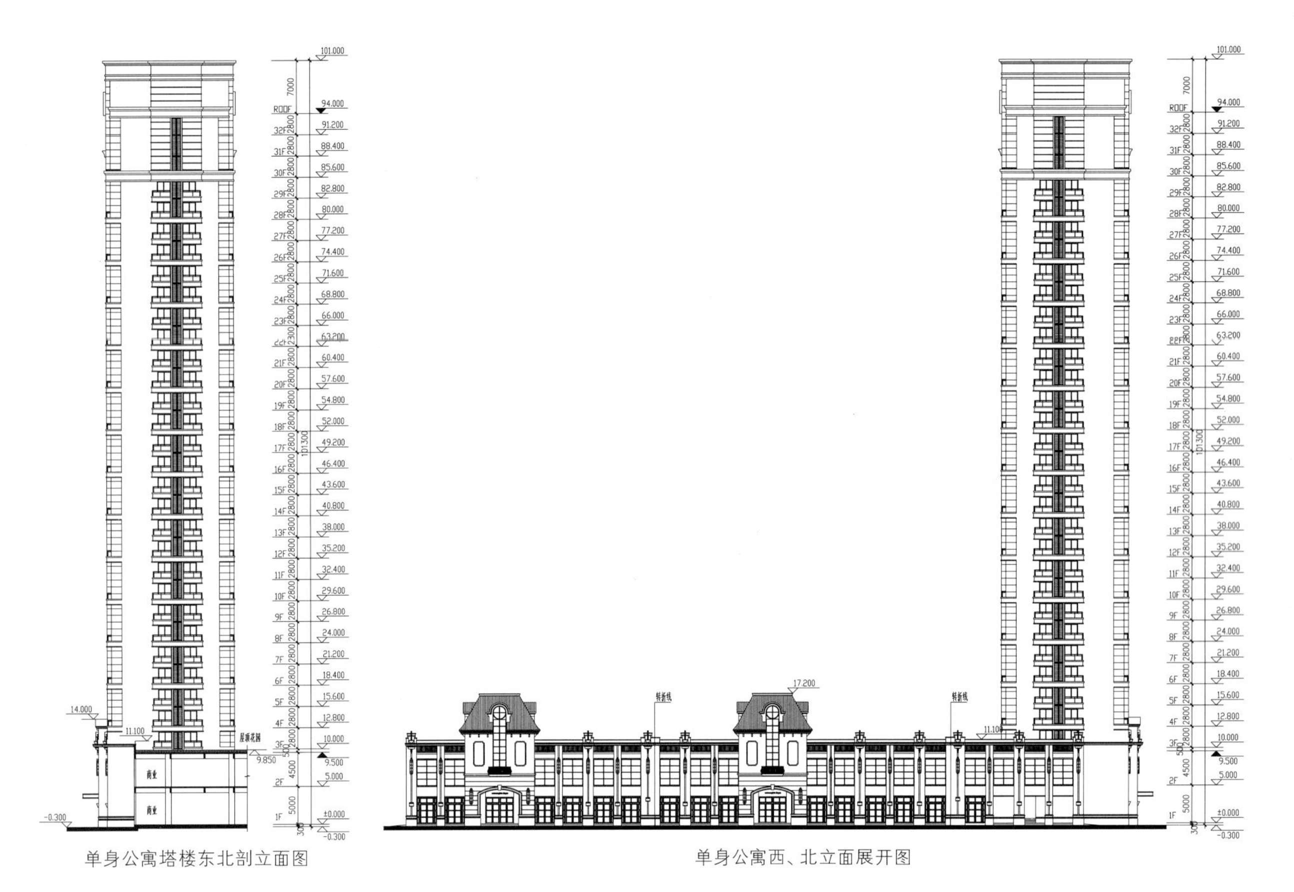

单身公寓塔楼东北剖立面图

单身公寓西、北立面展开图

转折线

94.900

101.000

7000

ROOF 94.000

32F 2800 91.200

31F 2800 88.400

30F 2800 85.600

29F 2800 82.800

28F 2800 80.000

27F 2800 77.200

26F 2800 74.400

25F 2800 71.600

24F 2800 68.800

23F 2800 66.000

22F 2800 63.200

21F 2800 60.400

20F 2800 57.600

19F 2800 54.800

18F 2800 52.000

17F 2800 49.200

16F 2800 46.400

15F 2800 43.600

14F 2800 40.800

13F 2800 38.000

12F 2800 35.200

11F 2800 32.400

10F 2800 29.600

9F 2800 26.800

8F 2800 24.000

7F 2800 21.200

6F 2800 18.400

5F 2800 15.600

4F 2800 12.800

3F 2800 10.000

500 9.500

2F 4500 5.000

1F 5000 ±0.000

300 -0.300

101.300

17.200

14.000

转折线

11.100

单身公寓东、南立面展开图

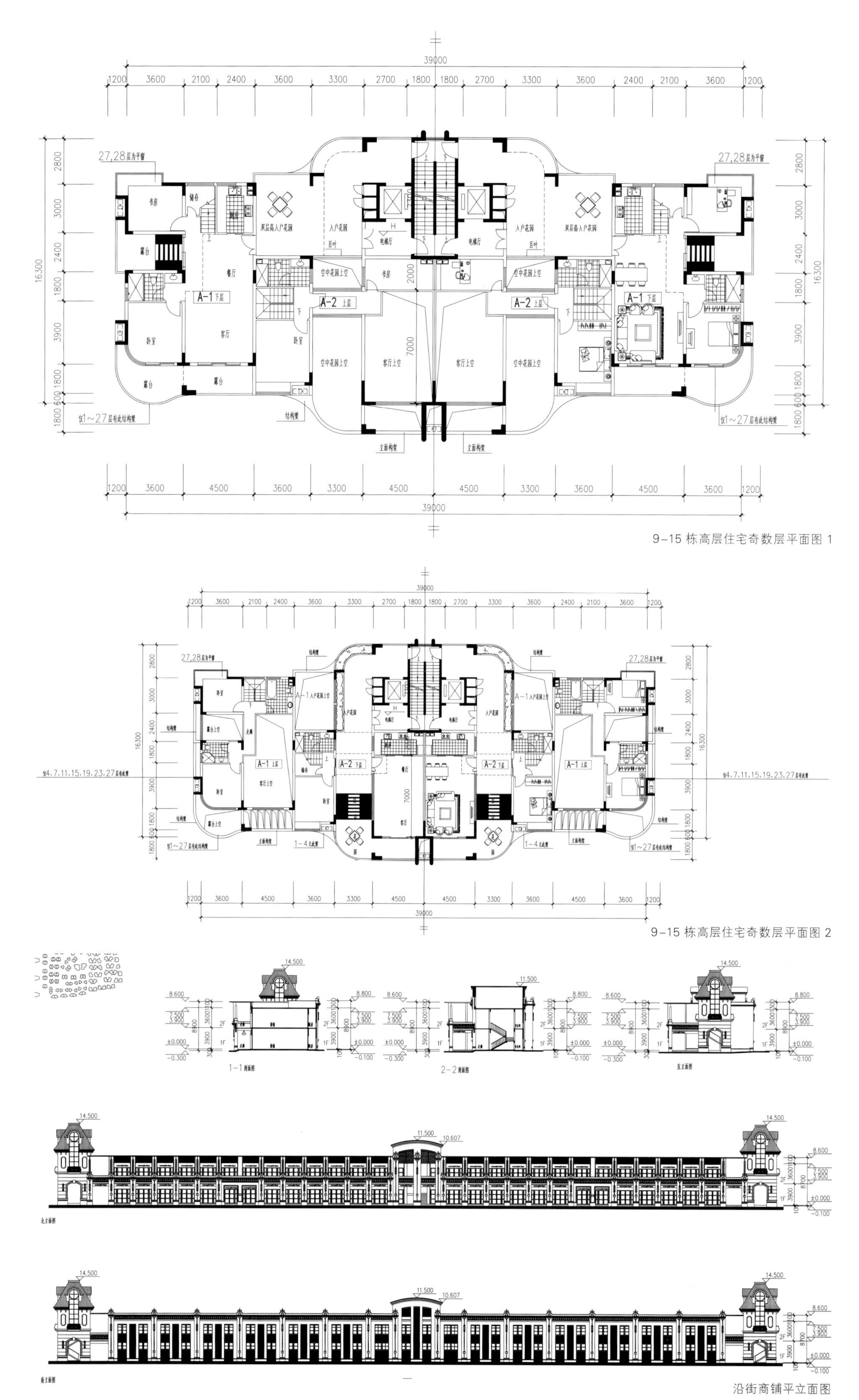

9-15 栋高层住宅奇数层平面图 1

9-15 栋高层住宅奇数层平面图 2

沿街商铺平立面图

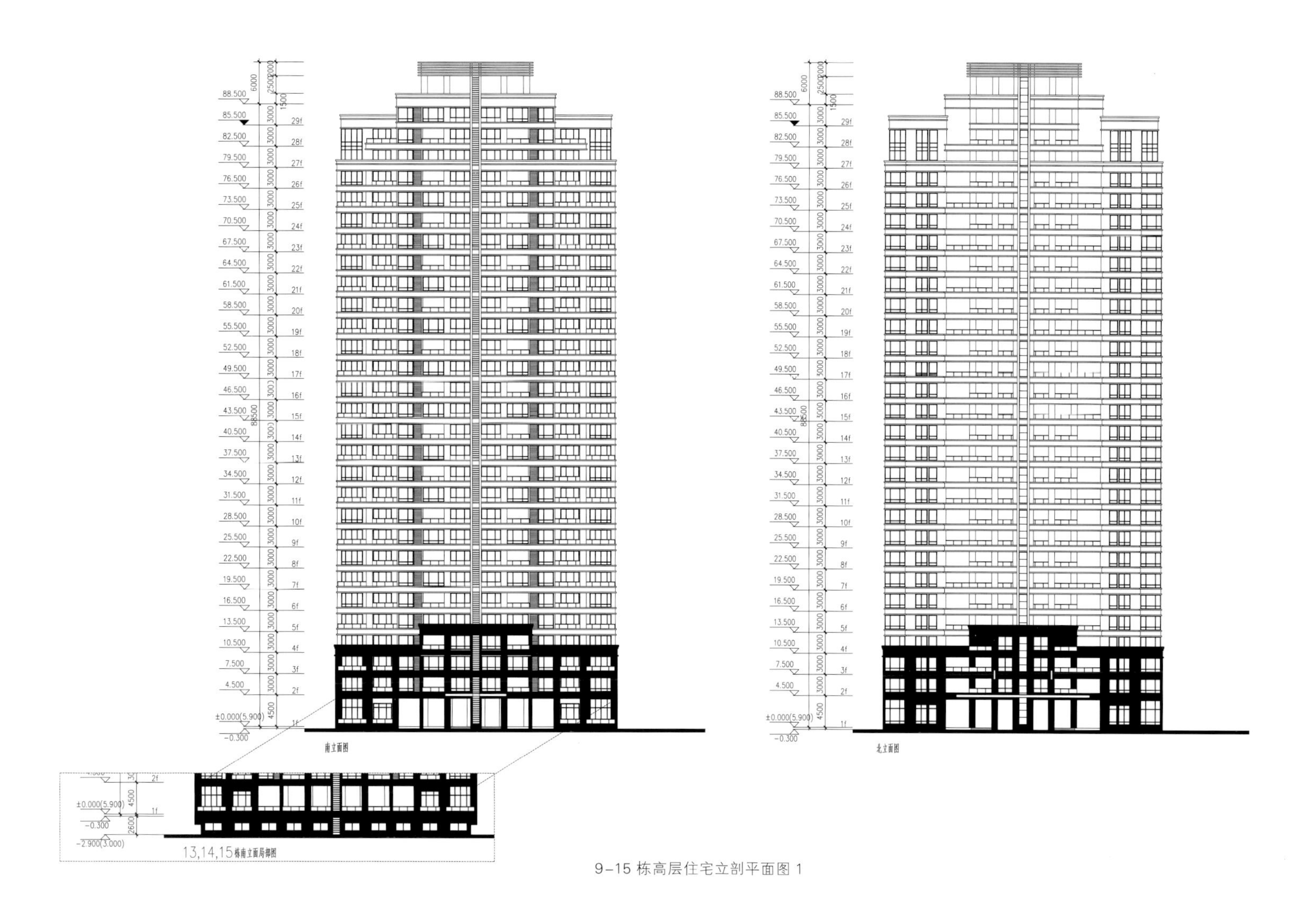

9-15 栋高层住宅立剖平面图 1

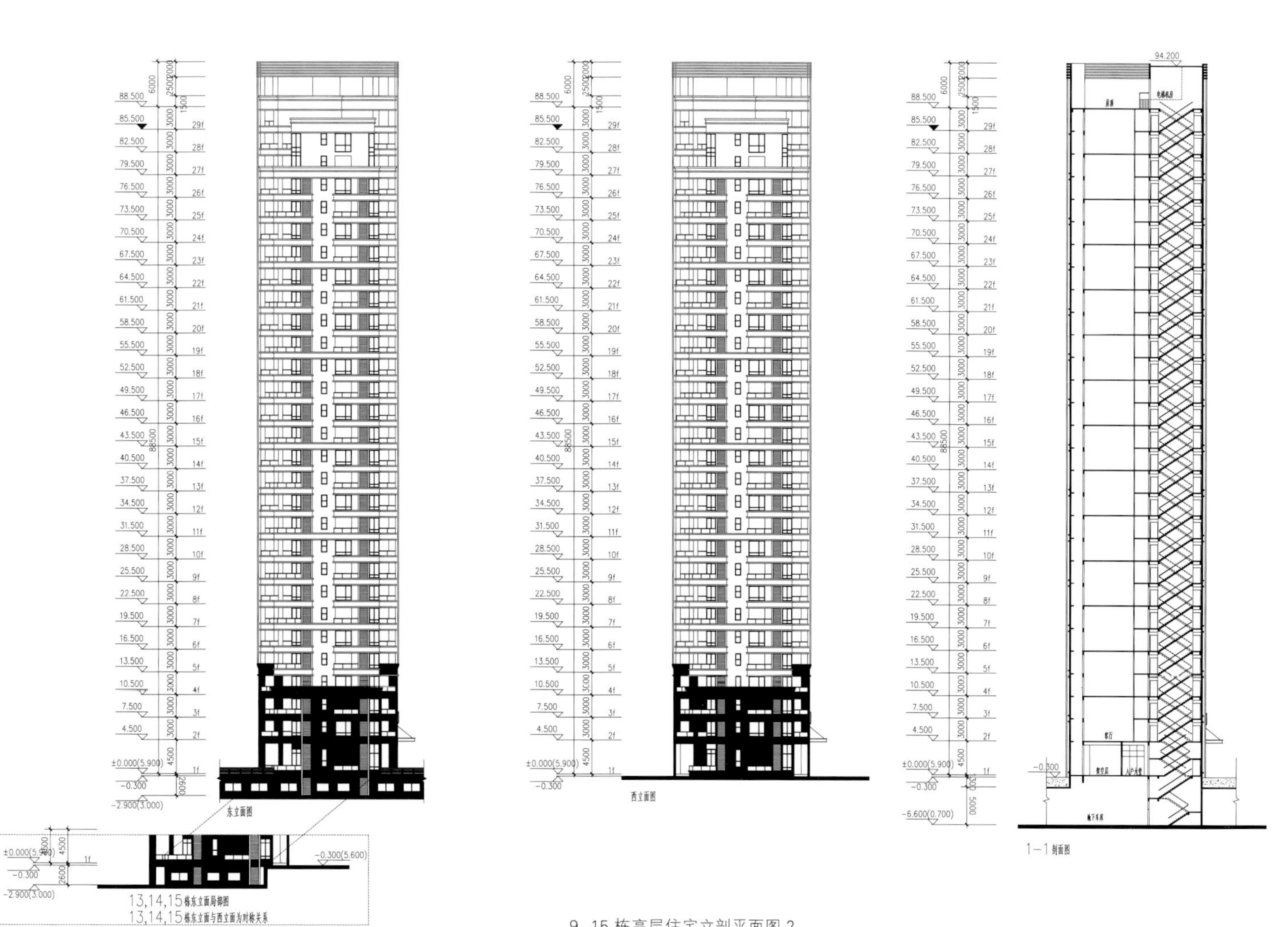

9-15 栋高层住宅立剖平面图 2

Royal Palace
尚观御园

Location: Foshan, Guangdong, China
Architectural Design: Shing & Partners
Land Area: 77,110.12 m^2
Floor Area: 199,145.3 m^2

项目地点：中国广东省佛山市
设计单位：汉森伯盛国际设计集团
占地面积：77 110.12 m^2
建筑面积：199 145.3 m^2

KEYWORDS 关键词

Every House Has Landscape
户户有景

Beautiful French Style
臻美法式

Humanization
人性化

FEATURES 项目亮点

This project takes beautiful French style architectures as the main buildings. The grand while not massive design style reflects a humanized ecological residential environment and high-quality living environment.

项目以臻美法式建筑为主体，大方而不厚重的设计风格，体现人性化的生态型居住环境和高品质的生活环境。

Overview

Royal Palace is located at Beihu First Road, Luocun Street, on the backbone network hinge of the planned "six lengthways roads, six transverse roads and one ring"; the traffic is very convenient with its north connecting to Shishan and Dali residential areas, south connecting to Chancheng central residential area, east to Guicheng and west to Danzao. This project takes beautiful French style architectures as the main buildings; the grand while not massive design style reflects a beautiful and precious living place.

项目概况

尚观御园，雄踞罗村街道北湖一路，坐落于规划建设中的“六纵六横一环”骨干路网的枢纽部位，北面接狮山组团、大沥组团，南面接通禅城中心组团，东临桂城，西连丹灶，交通非常便利。项目以臻美法式建筑为主体，大方而不厚重的设计风格，呈现一个美丽的人居宝地。

Design Principles

1. Culture, relaxation, and advocation of new living experience.

2. Every house has landscape which maximizes the value.

3. With wide distance from each other, the buildings form a more humanized living space.

4. From the city view, paying attention to the community image observed from outside, to make it a beautiful city scenery.

5. Promoting humanistic spirit, respecting the living requirements of residents, focusing on the living habits of residents, taking care of the daily life of residents, insisting on the principle of people orientation and fully combining the characteristics of the site to create a humanized ecological and high–quality living environment, and a humanistic and characteristic residential community with clear function, reasonable layout and rich spaces.

6. According to market requirements, reflecting the characteristics of the times, best meeting the needs of users, the design is pursuing the characteristics of practicalness, rationality and irreplaceability within certain limits. In this way it can also be beneficial to the site's integrated environment, service quality and social benefits.

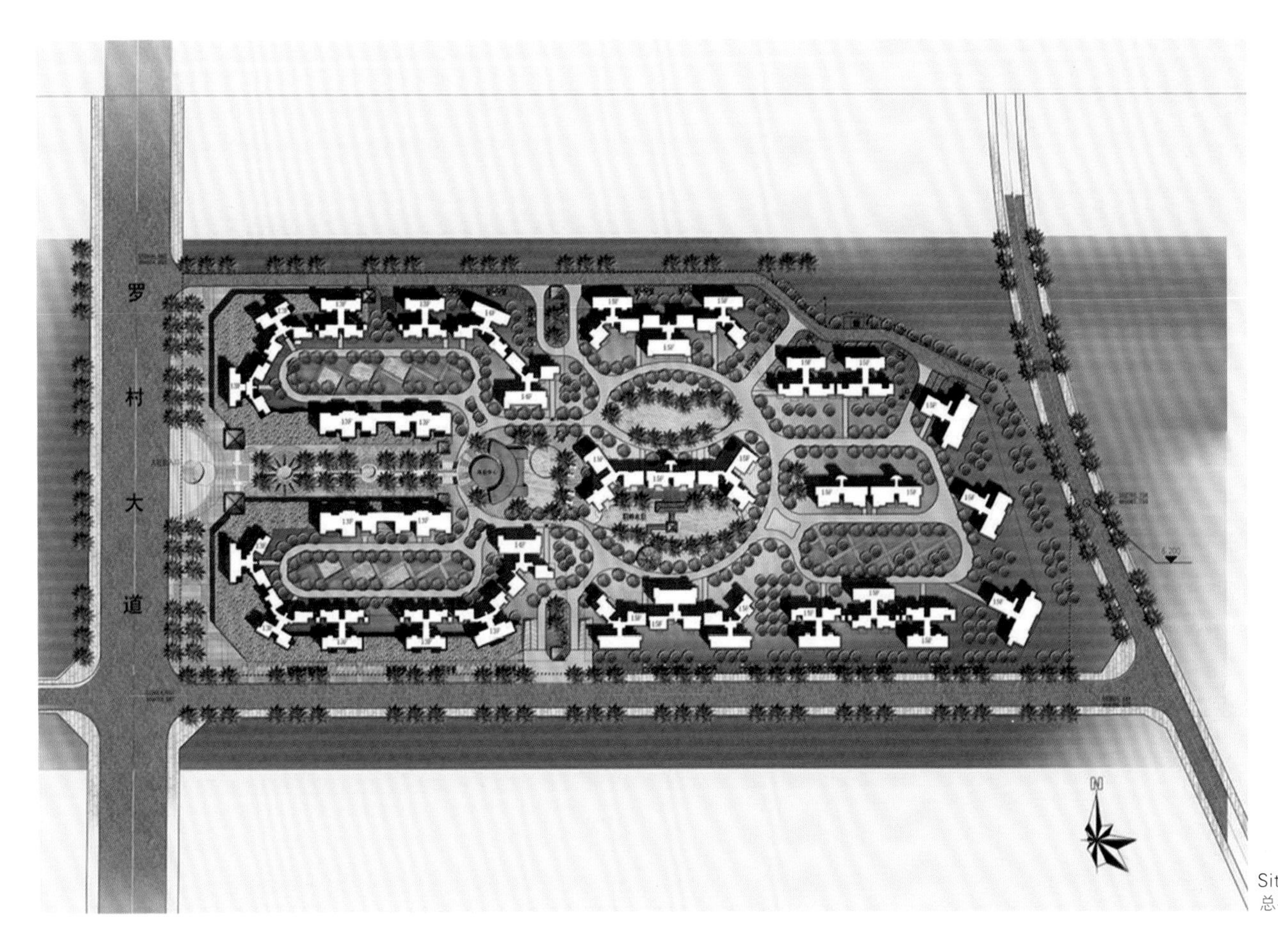

Site Plan
总平面图

➤ 设计原则

1. 文化、休闲、尊尚新型人居体验。

2. 户户有景，景观使价值最大化。

3. 宽大楼距，构筑更加人性化的生活空间。

4. 城市视角，重视从城市外围所看到的小区形象，还城市一道美丽的风景线。

5. 推崇人本主义精神，尊重住户的居住要求，关注住户的生活习惯，照顾住户的日常生活，坚持“以人为本”的设计原则，充分结合地块的特征，创造出人性化的生态型居住环境和高品质生活环境，创造一个功能明确、布局合理、空间丰富、既具有人文性又能充分发挥地块特色的住宅小区。

6. 针对市场需求，体现时代特征，最大限度地满足使用者的需要，在设计中追求实际、合理及在一定范围内不可替代的特点，帮助提高小区所在地的整体环境品位、服务质量和社会效益。

护泳池水质环境
泳帽着泳装 池

Qingdao Hisense Luxury Mansion
青岛海信君汇

Location: Qingdao, Shandong, China
Developer: Hisense Group
Architectural Design: W&R Group
Land Area: 56,200 m^2
Floor Area: 88,732 m^2
Plot Ratio: 0.55
Green Coverage Ratio: 40%

项目地点：中国山东省青岛市
开发商：海信集团
景观设计：水石国际
占地面积：56 200 m^2
建筑面积：88 732 m^2
容积率：0.55
绿化覆盖率：40%

KEYWORDS 关键词

Sufficient Sunshine
充足光照

Creating Micro Space
微空间营造

Echo Between Exterior and Interior
室内外呼应

FEATURES 项目亮点

Architects skillfully use the elements such as French columns, carving and lines for the details which are exquisite and elegant.

项目在细节处理上巧妙地运用了法式廊柱、雕花、线条等元素，精细而考究。

Overview

Luxury Mansion is developed as the high-end villas located on Mai Island with French architectural style facades. The landscape of project is defined as modern and natural cultural landscape, using modern natural landscape as the background of French and neoclassic architecture to highlight the dignity and taste of the buildings.

项目概况

青岛海信君汇为麦岛顶级高端别墅产品，别墅建筑为法式风格立面。设计将别墅环境定义为现代、自然的文化主题景观，现代自然的景观环境作为法式、新古典建筑的背景，凸显建筑主体的尊贵与品质。

Planning

There are 18 villas, a 7-storey and a 5-storey office buildings planned for the site, while the aboveground floor area of villas are ranging from 570 m^2 to 1,300 m^2, with four storeys in total and two storeys aboveground and underground respectively.

项目规划

该片区建设18栋独栋，1栋7F写字楼与1栋5F写字楼。别墅地上面积区间570～1 300 m^2，地上两层，地下两层设计。

Site Plan 总平面图

➤ Design Concept

Architects optimize the general layout and construct high-end space, adopting the detailed design methods including elevating the entrance, studying on the hallway, creating micro space, echoing between exterior and interior, as well as microtopography to build high quality environmental experience. With the introduction of culture and art themes to shape high-end taste for the villas, the environment of villas like an artistic manor with collection theme, also cooperating with decorations that seem made by hand to create unique and high-end atmosphere, moreover, the ways of gardening of soft landscape design are selected to build an artistic garden space.

➤ 设计理念

设计从优化规划布局到营造高品质空间，运用入口空间提升、入户空间研究、微空间营造、室内外呼应、微地形与空间等具体设计手法营造高品质化环境体验，通过文化艺术主题的导入塑造别墅区高端格调，别墅环境犹如进入一片收藏主题艺术庄园，并通过具手工感设计打造高端产品独一无二的质感，软景设计上运用“园艺化”手法营造精致的园林艺术化空间。

➤ Architectural Design

French style is adopted for the overall architectural style to manifest the noble flavor. The project is mainly developed with villas, emphasizing on the axis symmetry and grand atmosphere. Architects skillfully use the elements such as French columns, carving and lines for the details which are exquisite and elegant. Besides, in order to provide sufficient sunshine and delight for residents, French windows, dormers and large-sized windows are designed in the buildings flexibly.

➤ 建筑设计

项目的整体建筑风格为法式风格，彰显贵族风范。项目以独栋别墅为主，在布局上突出轴线的对称和恢宏的气势。在细节处理上巧妙地运用了法式廊柱、雕花、线条等元素，精细而考究。此外，为能使户主感受到充足的光照和生活情趣，落地玻璃、老虎窗以及大面积窗户都被灵活运用到建筑设计中去。

New Jiangwan Town Capital, Shanghai

上海新江湾城首府

Location: Yangpu District, Shanghai, China
Developer: Shanghai Chengtou Real Estate (Group) Co., Ltd.
Architectural Design: Shanghai ZF Architectural Design Co., Ltd.
Gross Land Area: 67,931.7 m^2
Gross Floor Area: 151,791.01 m^2

项目地点：中国上海市杨浦区
开发商：上海城投置地（集团）有限公司
设计单位：上海中房建筑设计有限公司
总占地面积：67 931.7 m^2
总建筑面积：151 791.01 m^2

KEYWORDS 关键词

Slope Roof
坡顶

Cast Iron Forging
铸铁锻造

Sequential Space
礼序空间

FEATURES 项目亮点

Based on the theme of "French Chateau", the residential buildings are designed with classical French-style facades. From site planning, architectural design and landscape design to interior decoration, they all contribute to a pure French architectural style.

基于"法式红酒庄园"的主题定位，住宅建筑立面采用古典法式建筑风格，从规划设计、建筑设计、景观设计到室内精装，营造纯正法式建筑风格。

Overview

Located in the center of New Jiangwan Town, Yangpu District, Shanghai, the site is near to the Ecological Wetland Park on the west; the newly-built Community Neighborhood Center and Shanghai Talent Apartment are on the north; to the south, the commercial center of New Jiangwan Town will be completed soon. Enjoying convenient supporting living facilities, the project will be another landmark of this area upon completion.

项目概况

项目位于上海杨浦区新江湾城核心区域，地块西临新江湾城生态湿地公园，北靠即将投入使用的社区邻里中心及上海市人才公寓，南侧为即将建成的新江湾城商业中心，生活配套设施极为便利。项目建成后，将成为新江湾城板块又一标杆性楼盘。

Architectural Design

The facades are designed in classical French style to match the theme of "French Chateau". With proper proportion and varied high-class materials (i.e., stone, metal, tiles), it has created a series of luxury and high-quality buildings. All these buildings are designed with slope roofs and different heights to form a dynamic skyline for the city, which are harmonious with the entire scenic area. The facades of the clubhouse and the public facilities are designed in the same style to highlight the theme of the community.

建筑设计

住宅建筑立面采用古典法式建筑风格，以体现业主对整体小区的"法式红酒庄园"主题的定位。通过对形体尺度的推敲，多种高档材质（如石材、金属、面砖等）的配合运用，形成高档而精致的品质感。建筑均采用坡顶，并通过高低错落的建筑布局形成较活泼的城市第五立面，与整个风景区相协调。会所与公建的立面设计也与整体小区立面相协调，烘托出完整的"法式红酒庄园"的社区氛围。

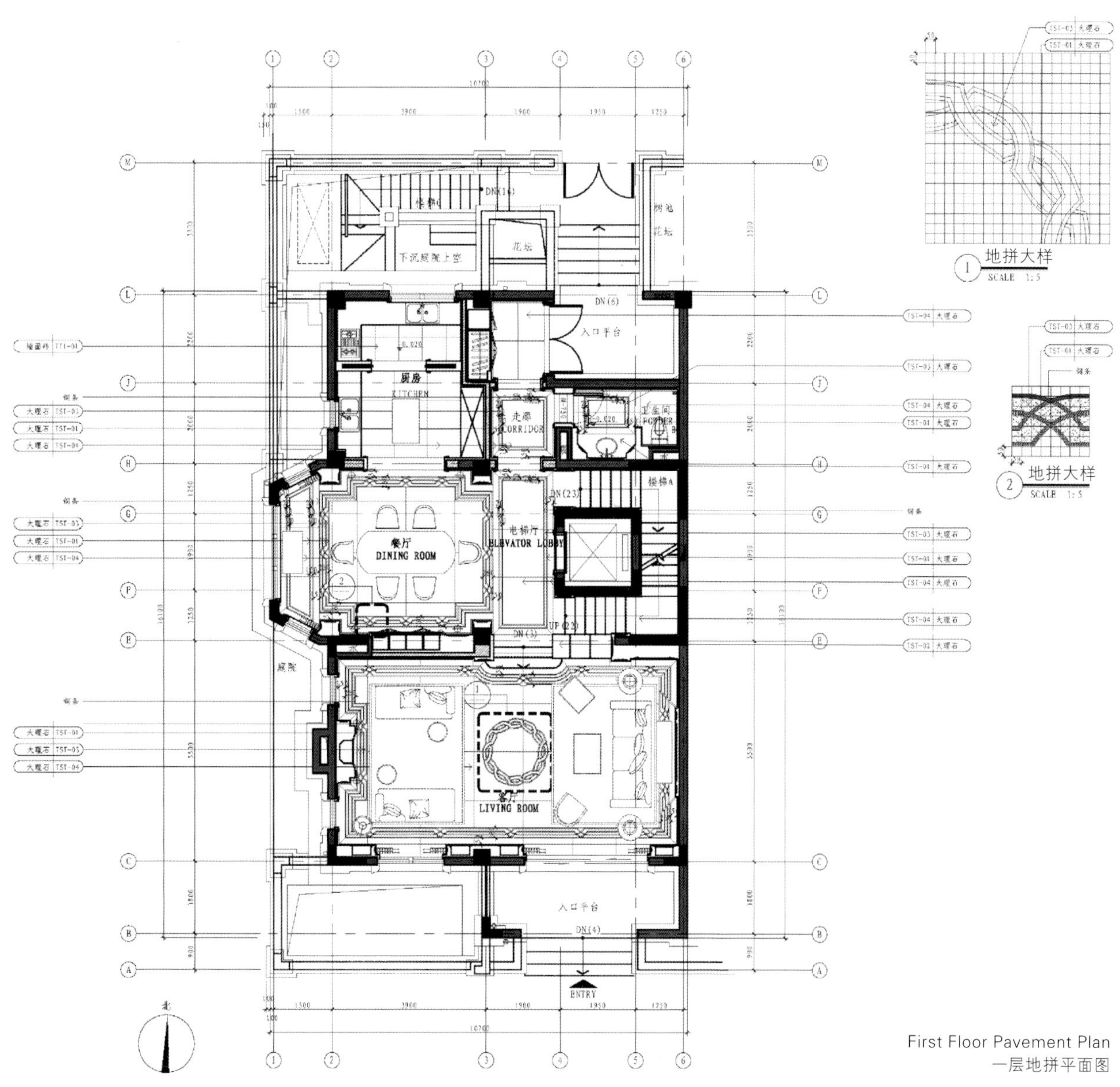

First Floor Pavement Plan
一层地拼平面图

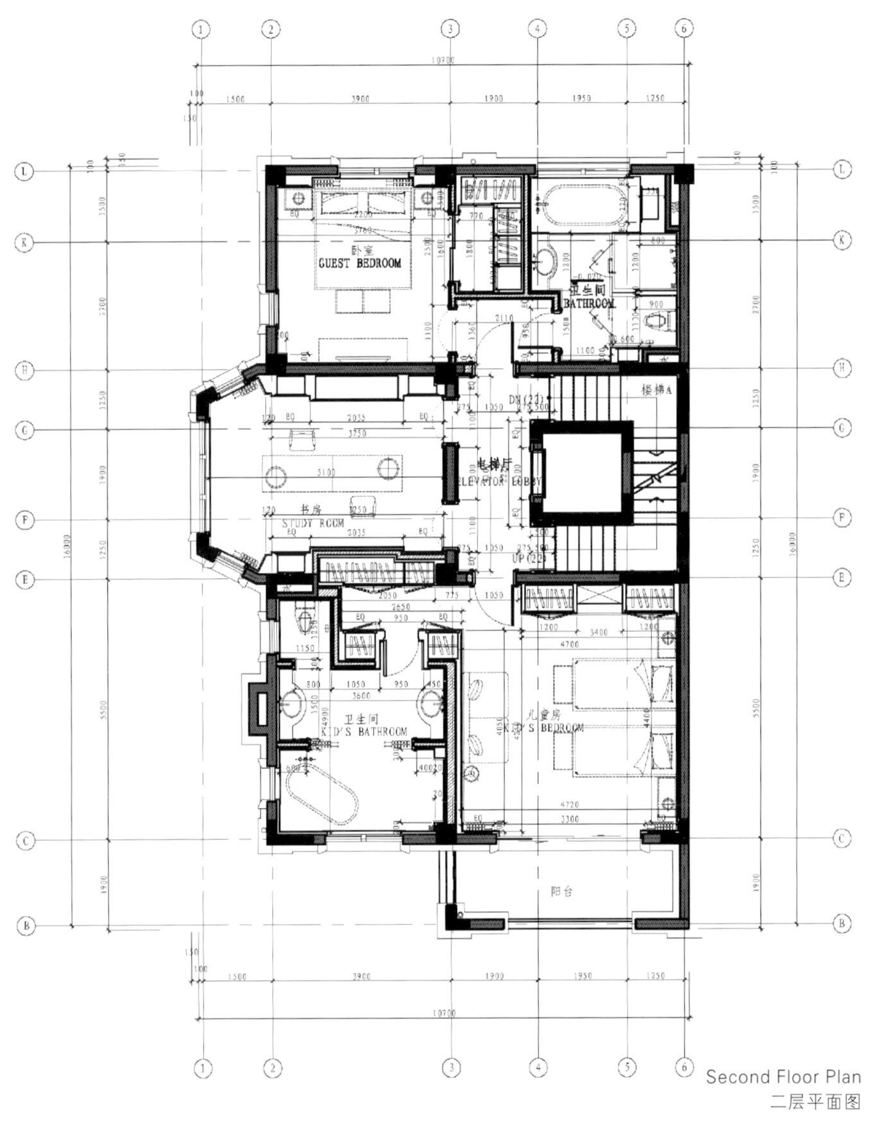

Second Floor Plan
二层平面图

➤ House Type Design

The plan design not only meets the functional requirements but also matches with the interior style that caters to the specific purchasing group. From the width and depth of the rooms, to the control of the areas, all the bedrooms are designed in suite standard. The arrangement of the public and private spaces, and the separation of the maters' and the housekeeper's traffic lines will enable the residents to enjoy a high-quality lifestyle.

➤ 户型设计

住宅平面设计在满足房型的功能性要求外，从对房间面宽进深的尺度推敲，到对户型面积的精细控制，实现所有的卧室均为全套间设计。公共空间与休息空间的动静流线的分离，主人与家政的交通流线的完全分离等人性化高品质设计，使居住者能清晰体会和享受高品质的生活和先进的家居理念。

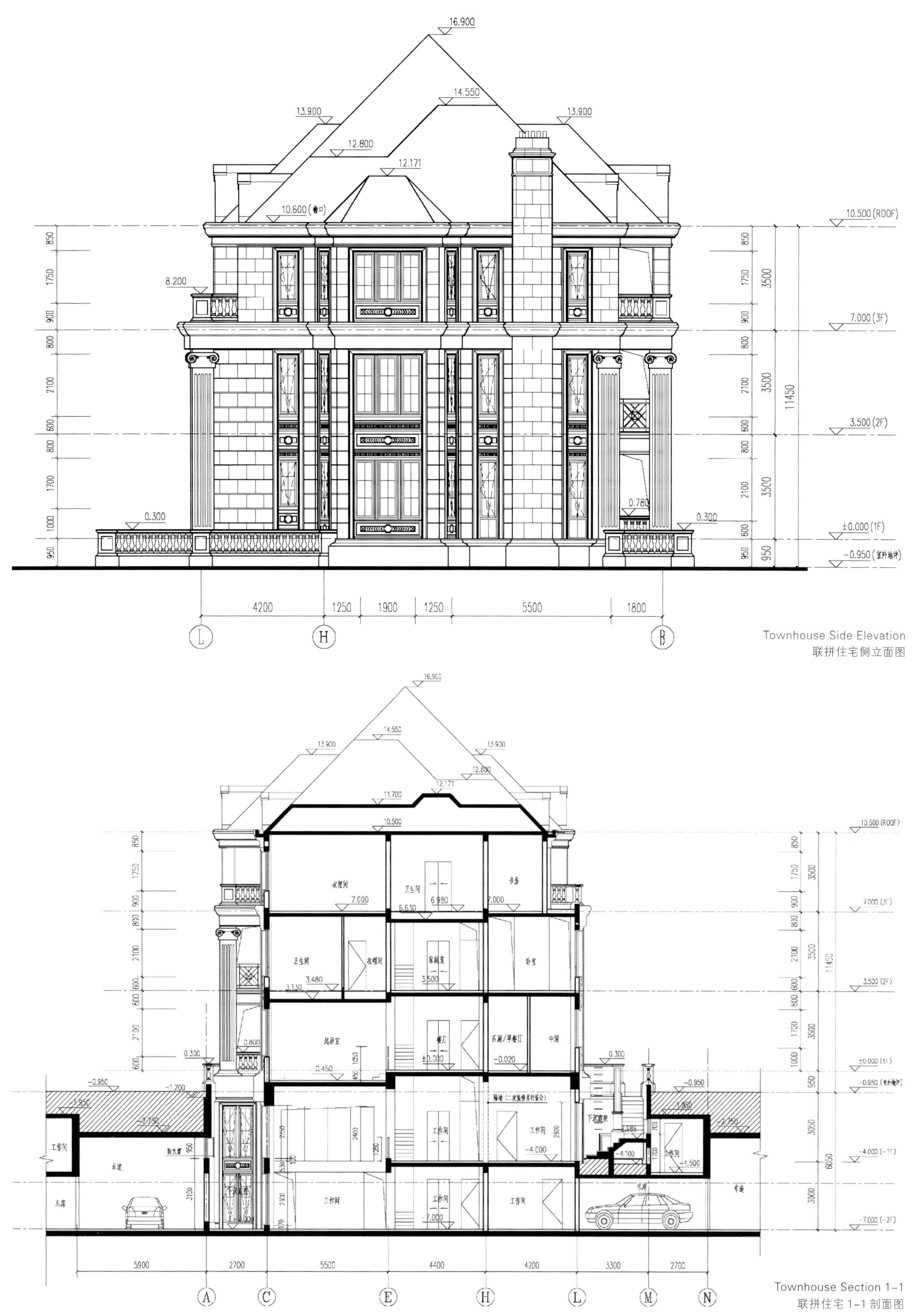

Townhouse Side Elevation
联拼住宅侧立面图

Townhouse Section 1–1
联拼住宅 1–1 剖面图

➤ Landscape Design

The landscape is designed in French palace and garden style to match the architectures. Landscape spaces are arranged in strict sequence. There are orderly and double axes. The south-north axis starts from the south entry (marble courtyard), and extends along the clubhouse, embroidery flower bed, rectangular lawn, pool and the Fountain of Neptune. It also connects with the riverfront landscape to provide more public spaces. The east-west landscape axis refers to the royal avenue intersecting the other axis and rolling with the topography. Along this axis, the trees and gardens are well designed to present a series of sequential spaces which are balanced and pleasant.

Four-unit Townhouses South Elevation
四联拼住宅南面图

▶ 景观设计

景观设计采取与规划建筑风格相一致的法式宫廷园林风格，严格遵循“礼序空间”的设计精髓，采用有规则的双轴线方案。南北主轴注重空间序列的庄重礼仪性，从小区南入口的“大理石庭院”开始，依次经过宏伟的会所、刺绣花坛、矩形草地、水池直到海神喷泉，并与河道沿河景观绿化相连通，增加了空间的共享性。东西景观轴是皇家林荫大道，与南北主轴相交，随着地势的起伏充满变化，轴线两侧的丛林、园艺则与轴线形成明暗、动静的对比，达到一种令人信服的动态均衡。

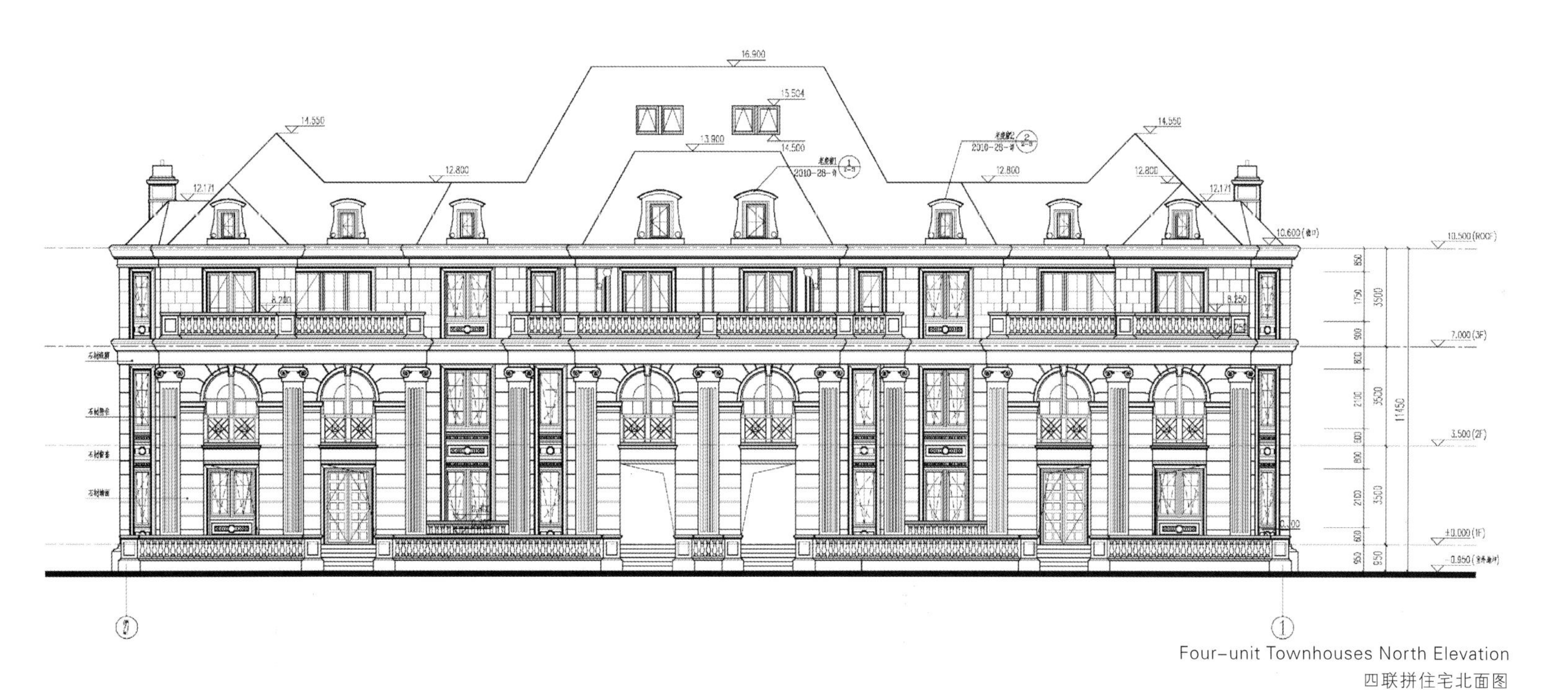

Four-unit Townhouses North Elevation
四联拼住宅北面图

NEW
JIANGWAN
CITY

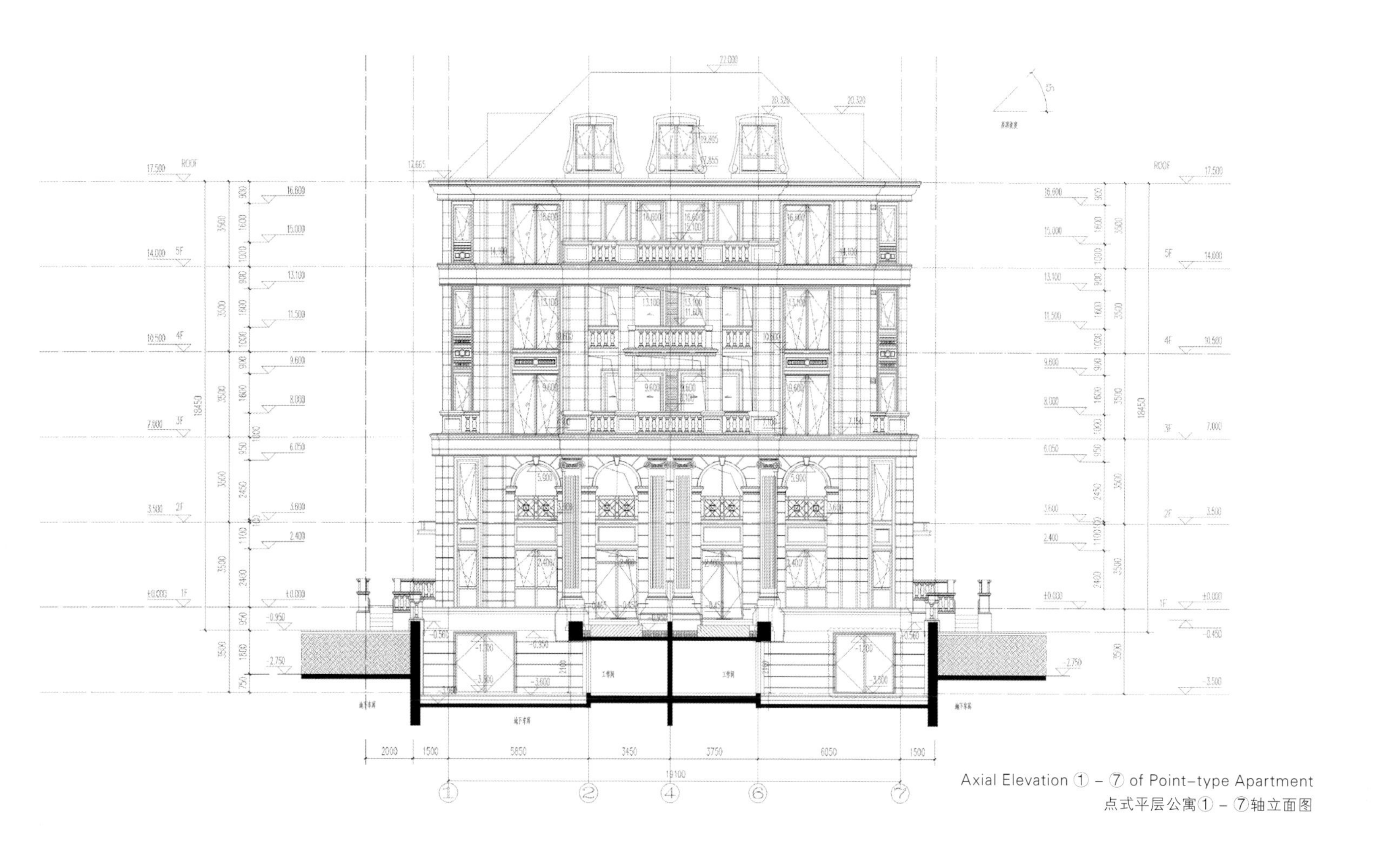

Axial Elevation ① – ⑦ of Point-type Apartment
点式平层公寓① – ⑦轴立面图

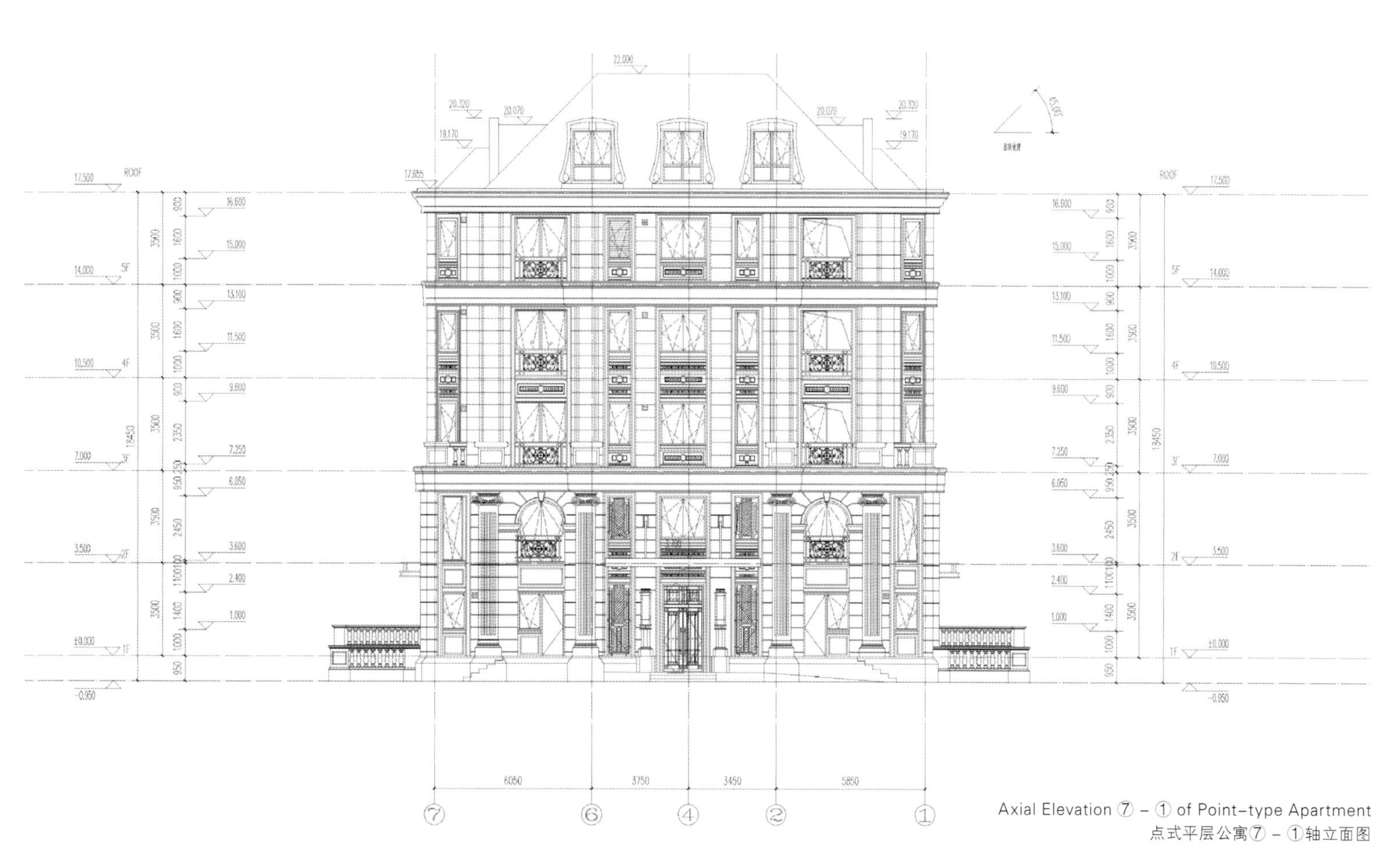

Axial Elevation ⑦ – ① of Point-type Apartment
点式平层公寓⑦ – ①轴立面图

Other Styles

其他风格

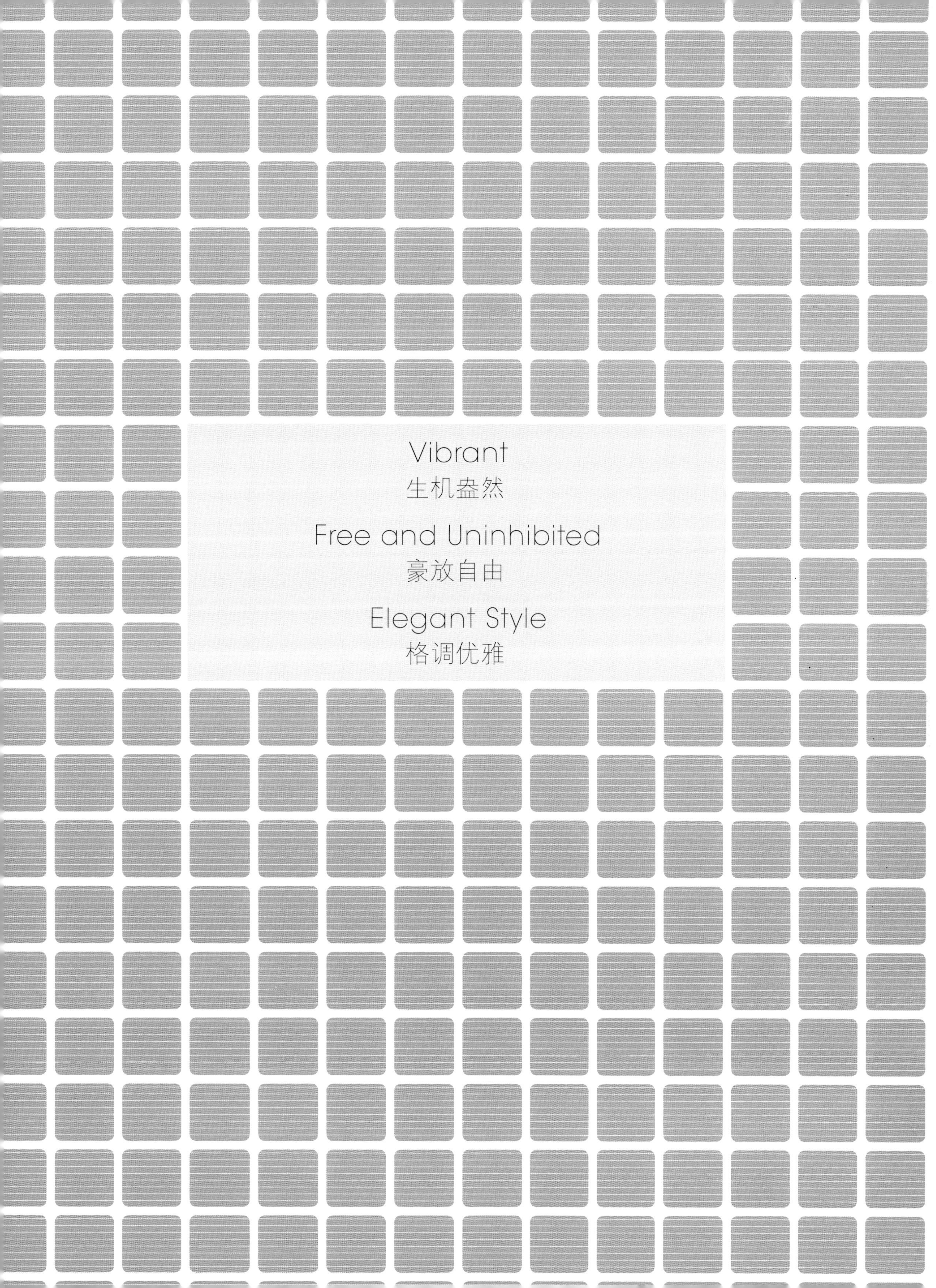

Vibrant
生机盎然
Free and Uninhibited
豪放自由
Elegant Style
格调优雅

Changzhou Longfor Project
常州龙湖香醍漫步

Location: Changzhou, Jiangsu, China
Developer: Longfor Property
Planning/Architectural Design: Shanghai Hyp-arch Architectural Design Consultant Inc.
Total Land Area: 308,820 m^2
Total Floor Area: 735,400 m^2

项目地点：中国江苏省常州市
开发商：龙湖地产
规划/建筑设计：上海霍普建筑设计事务所有限公司
总占地面积：308 820 m^2
总建筑面积：735 400 m^2

KEYWORDS 关键词

Accessible Scale
近人尺度

Concise Forms
形式简洁

Delicate Details
细部精致

FEATURES 项目亮点

In the architectural modelling design, the architectural form design focuses on architecture's humanization, expresses the meaning of "family", emphasizes on combination of modern and affection, while highlights the difference between this design and current popular form to reflect product's diversity and makes this residential form become the advanced classic project among future domestic residential design.

建筑造型设计中，在建筑形式设计中注重了建筑人性化，设计"家"的意义表达的建筑形式，侧重设计的现代性和情感性的结合，同时突出本设计与目前流行样式的差别，体现产品的差异性，使得该住宅形式成为未来国内住宅产品设计中的前沿经典之作。

➤ Overview

The project is located on the northwest of downtown in Changzhou City, west to Qingfeng Road, north to North Tanghe Road, east to Sanxin Road and south to Qingye Road, with Hejiatang River and Yongning Road running through. The base is adjacent to the East Longitude 120 Park, 1,500 m distance to Dinosaur Park, 4,000 m to New North District Center, 2,500 m to new Changzhou administrative center, 5,000 km to Changzhou Downtown Area (business center), with superior condition and convenient traffic.

➤ 项目概况

项目位居常州市中心城区西北部，西至青丰路，北至北塘河路，东至三新路，南邻青业路。贺家塘河及永宁路贯穿其中。基地毗邻规划中的东经120公园，距恐龙园1 500 m，新北区中心4 km，常州市新的行政中心2.5 km，常州市中心城区（核心商圈）5 km，区位条件优越、交通方便。

➤ Project Planning

The entire project planning is developed in three phases and all is low-rise and high-rise residential products. Phase I has finished the building of low-rise residence while the high-rise part is in building. Thus partly view can be presented.

The project planning starts with fluency, sharing and harmony. The overall layout puts high-rise residence surround outside of villa. Different with traditional approach of complete separation of high-rise area and villa area, this project will extend the high-rise area into villa area by open neighborhood space, link the axis of villa area and water landscape belt, maximize the value of high-rise dwelling and enhance the privacy through small node in villa area. Meanwhile the high-end residence developing experience of Longfor Property is incorporated to make delicate and smooth space.

➤ 项目规划

整个项目工程规划分三期开发。均为低层及高层住宅产品。目前在建一期开发项目，其中低层住宅均已交付，高层在建。部分实景已呈现。

项目规划将流畅、共享及和谐作为设计出发点。整体布局将高层呈围合状布置于别墅外围，有别于传统将高层区与别墅社区完全分隔的手法，本项目将开放性邻里空间作为纽带，将高层的活动范围延伸至别墅区，连接别墅区主轴与亲水景观带，最大化高层居住价值，并通过设置别墅小尺度节点加强私密性。同时融入龙湖地产多年高端住宅开发经验，整体氛围一气呵成，空间精致流畅。

Site Plan
总平面图

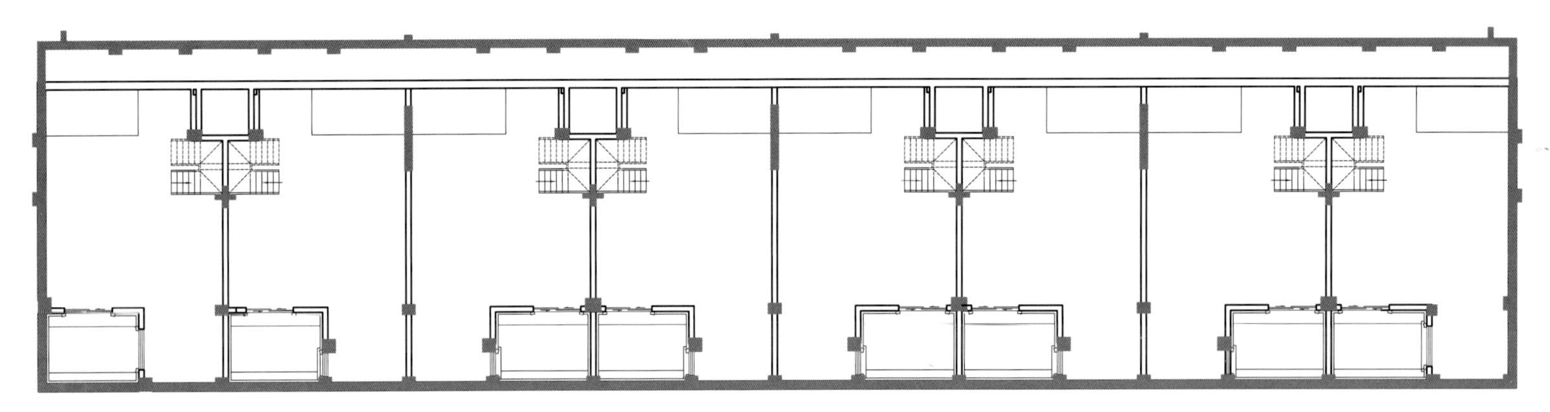

Basement Floor Plan 地下室平面图

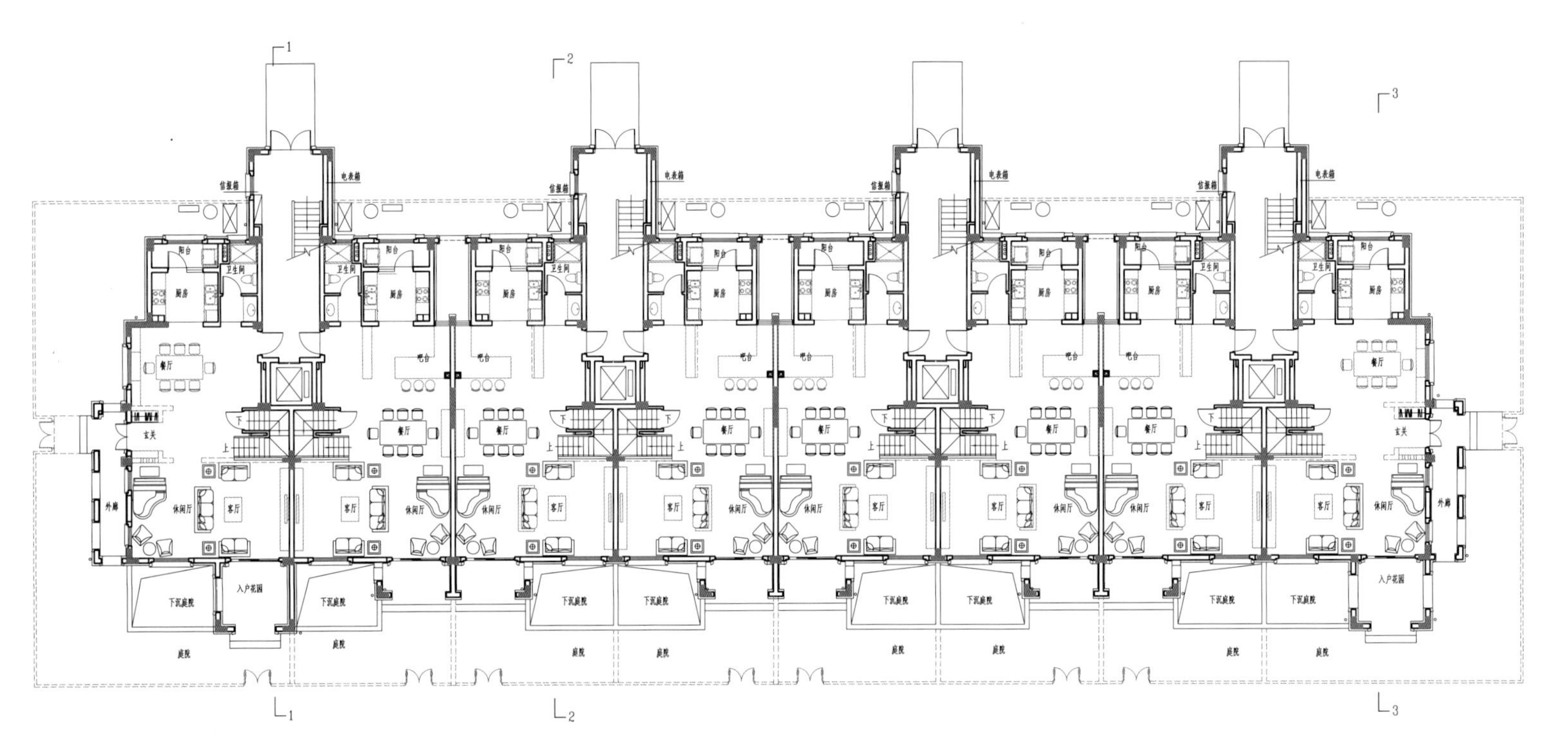

First Floor Plan 一层平面图

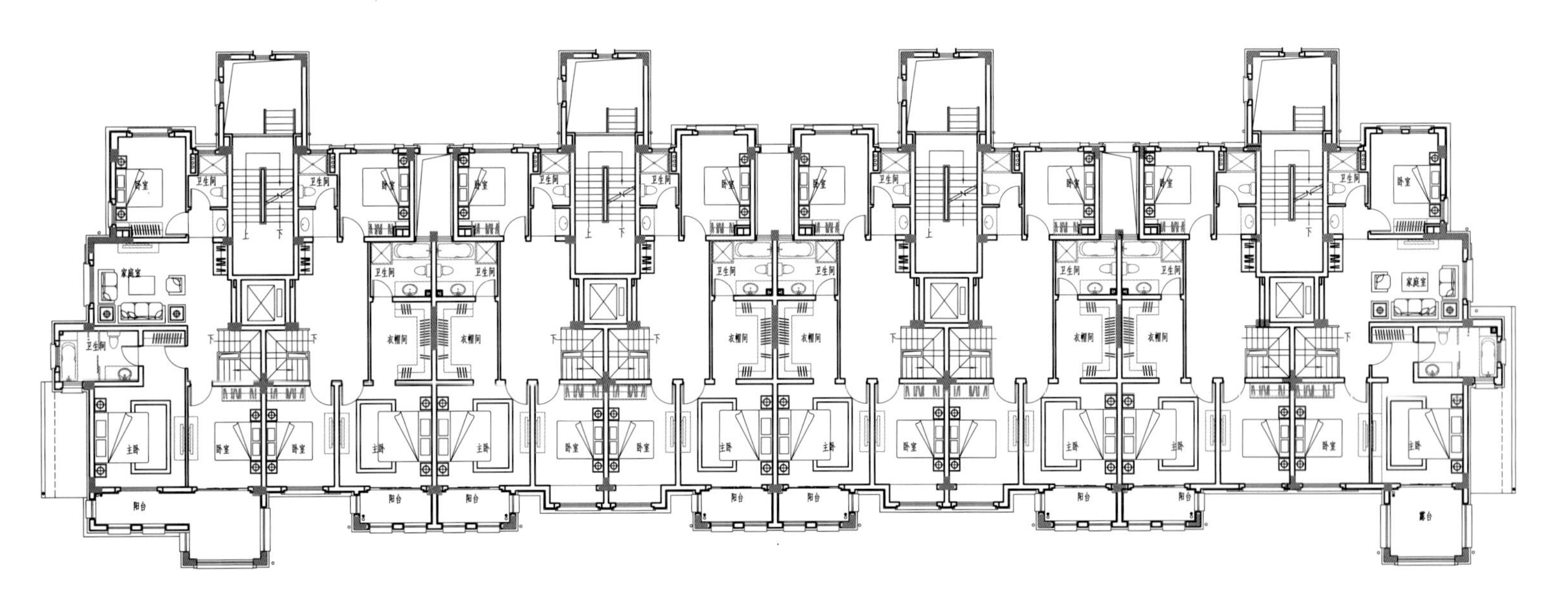

Second Floor Plan 二层平面图

The townhouse mainly is 220 m^2 and its different architectural styles provide rich choices for owners' different need. The housing design with reasonable function division satisfies owners' living requirements. House's inside centers on living rooms, with clear separation of privacy and public, living and sleeping area, cleanness and dirt. Compact interior layout and short passage improve the area's usage rate and comfort degree. High-rise residential type design on the basis of the overall layout is flexible and changeable and creates more interior space for natural ventilation and lighting to save energy. The plane division is clear and fluent, with economic and reasonable layout.

Residential flat emphasizes the design principles of natural lighting and ventilation for hall, kitchen and washroom. Hall is square and practical, uniformly equipped with outdoor air conditioner bit, condensate pipes, kitchen flue, water heaters air outlet and etc. Each household has work balcony, and the main balcony's depth is not less than 1.5 m, with clothesline facilities. Water, electricity and gas meters are out of the house. Residential area has underground garage. Single design emphasizes building's accessible scale, thus spatial variation and form treatment of detail design and end unit make full use of landscape resources. In the architectural modelling design, the architectural form design focuses on architecture's humanization, expresses the meaning of "family", emphasizes on combination of modern and affection, while highlights the difference between this design and current popular form to reflect product's diversity and make this residential form become the advanced classic project among future domestic residential design. Design uses concise forms, delicate details as well as modern and vibrant colors to embody future trends in residential architecture style and create unique personality traits.

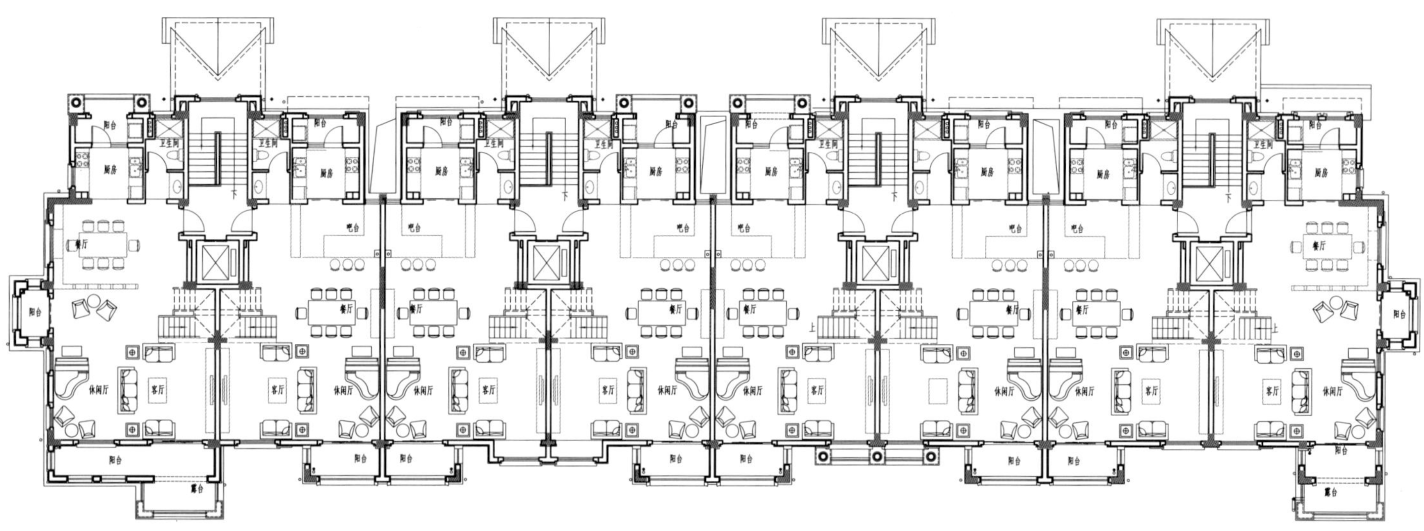

Third Floor Plan 三层平面图

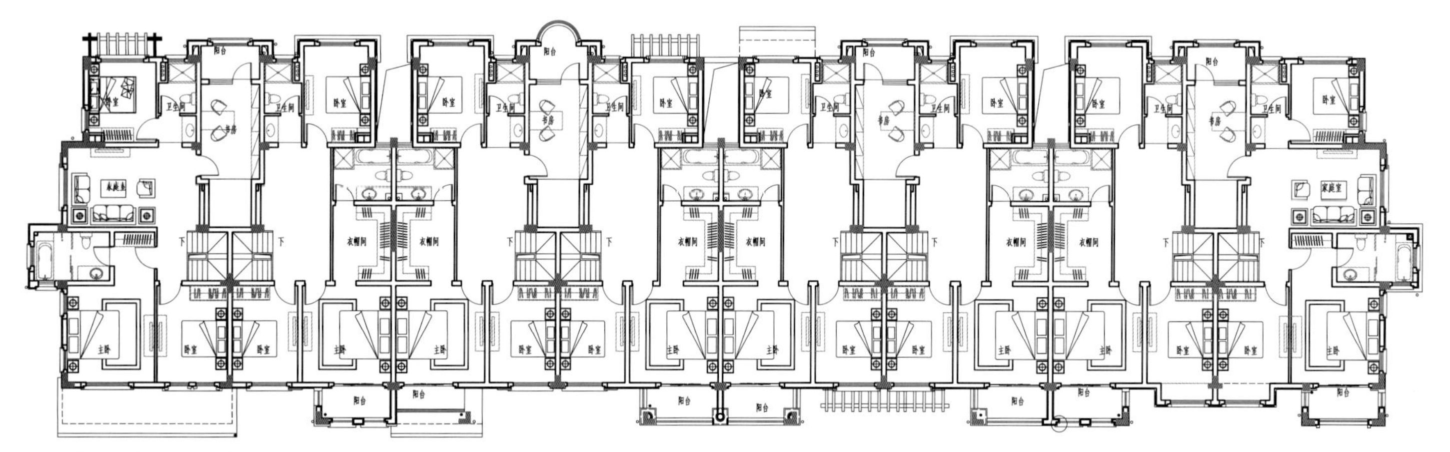

Fourth Floor Plan 四层平面图

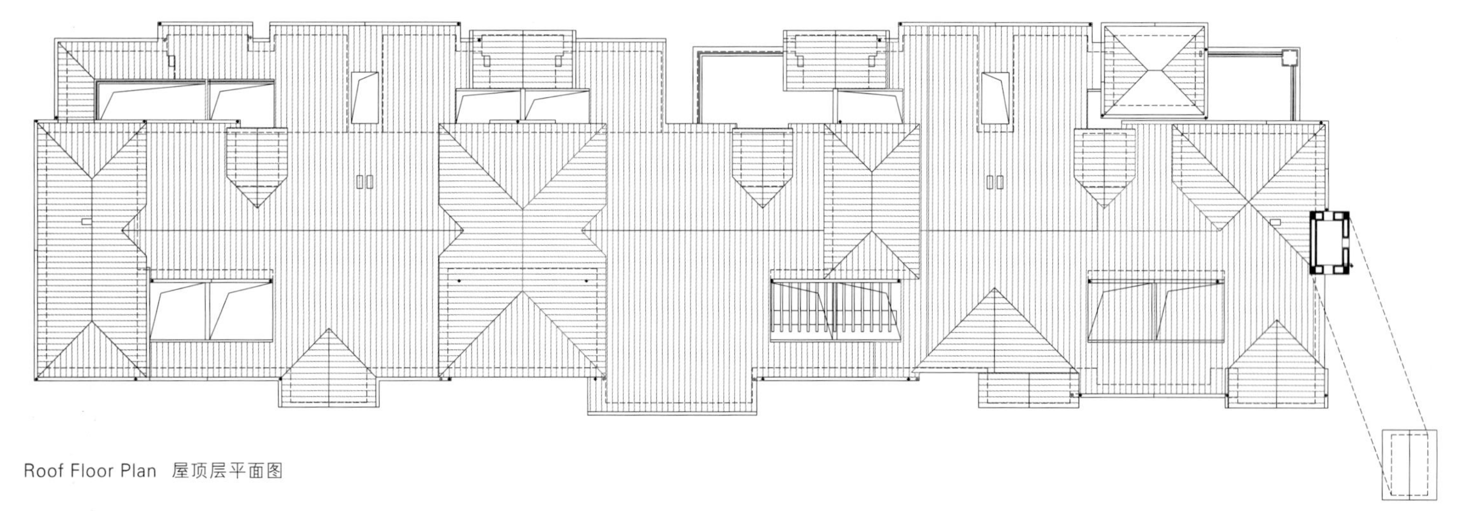

Roof Floor Plan 屋顶层平面图

联排别墅以 220 m^2为主，不同的建筑风格给有不同需求的业主提供丰富的选择。功能分区合理的房型设计满足了业主的生活需要，住宅套内设计以起居室为中心，内部空间公私分离、居寝分离、洁污分离、室内布置紧凑，走道短捷，提高面积的使用率和使用的舒适程度。高层住宅套型设计依据总体布局灵活多变，创造更多的自然通风、采光的室内空间，节约能源。平面分区明确，流线顺畅，布局经济合理。

住宅平面强调明厅、明厨、明卫及自然采光与通风的设计原则，厅房方正实用，统一设计室外空调机位、冷凝水管、厨房烟道、热水器排气口等，每户皆有工作阳台，主阳台进深不少于1.5 m，并有晒衣设施。水、电、煤表均出户。住宅设地下车库。单体设计强调建筑近人尺度，细部设计以及尽端单元的空间变化和形式处理，充分利用景观资源。建筑造型设计中，在建筑形式设计中注重了建筑人性化，设计“家”的意义表达的建筑形式，侧重设计的现代性和情感性的结合，同时突出本设计与目前流行样式的差别，体现产品的差异性，使得该住宅形式成为未来国内住宅产品设计中的前沿经典之作。设计中以简洁的形式、精致的细部，以及现代而鲜明的色彩来体现未来住宅建筑风格的趋势，塑造独特的个性特征。

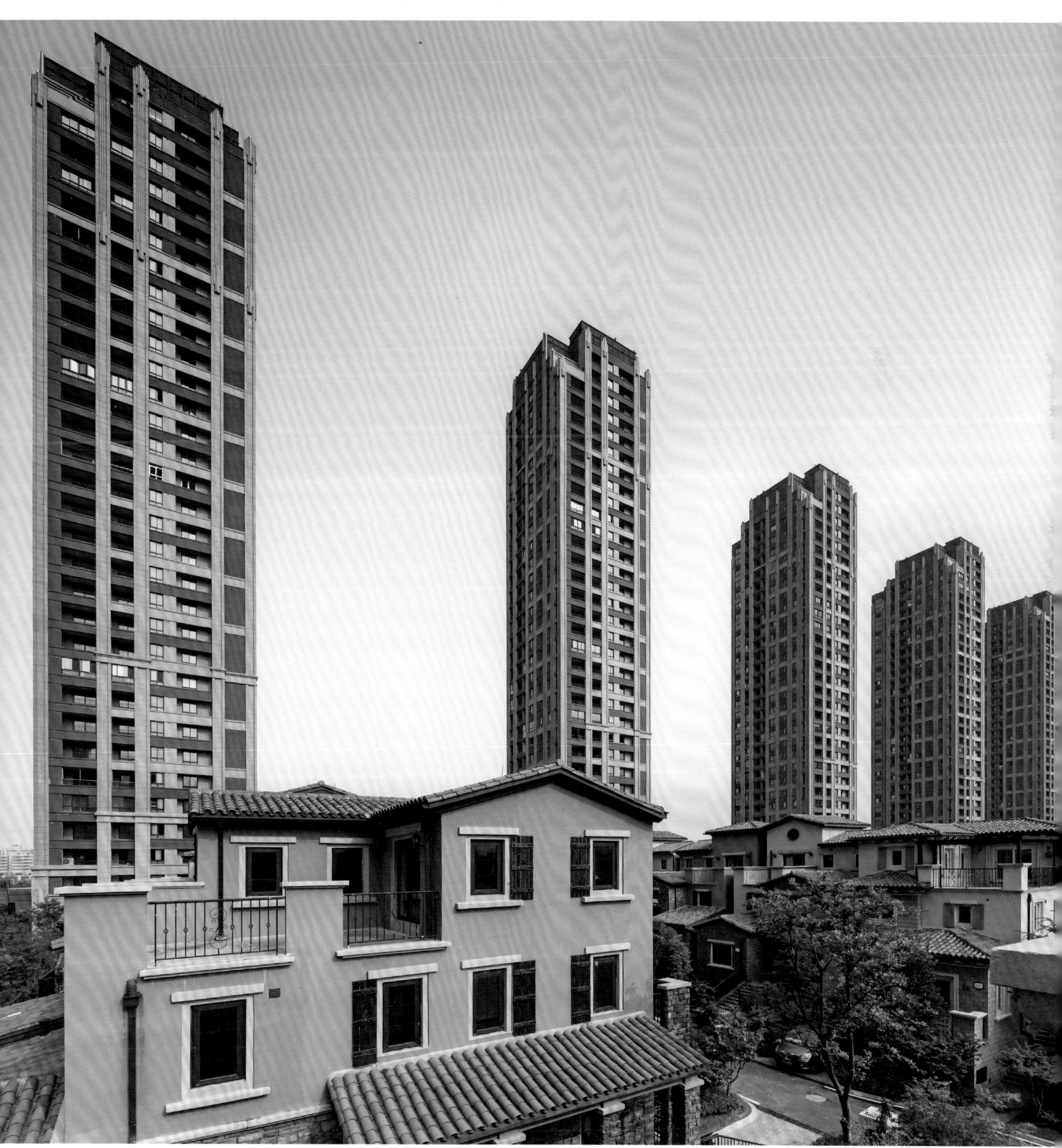

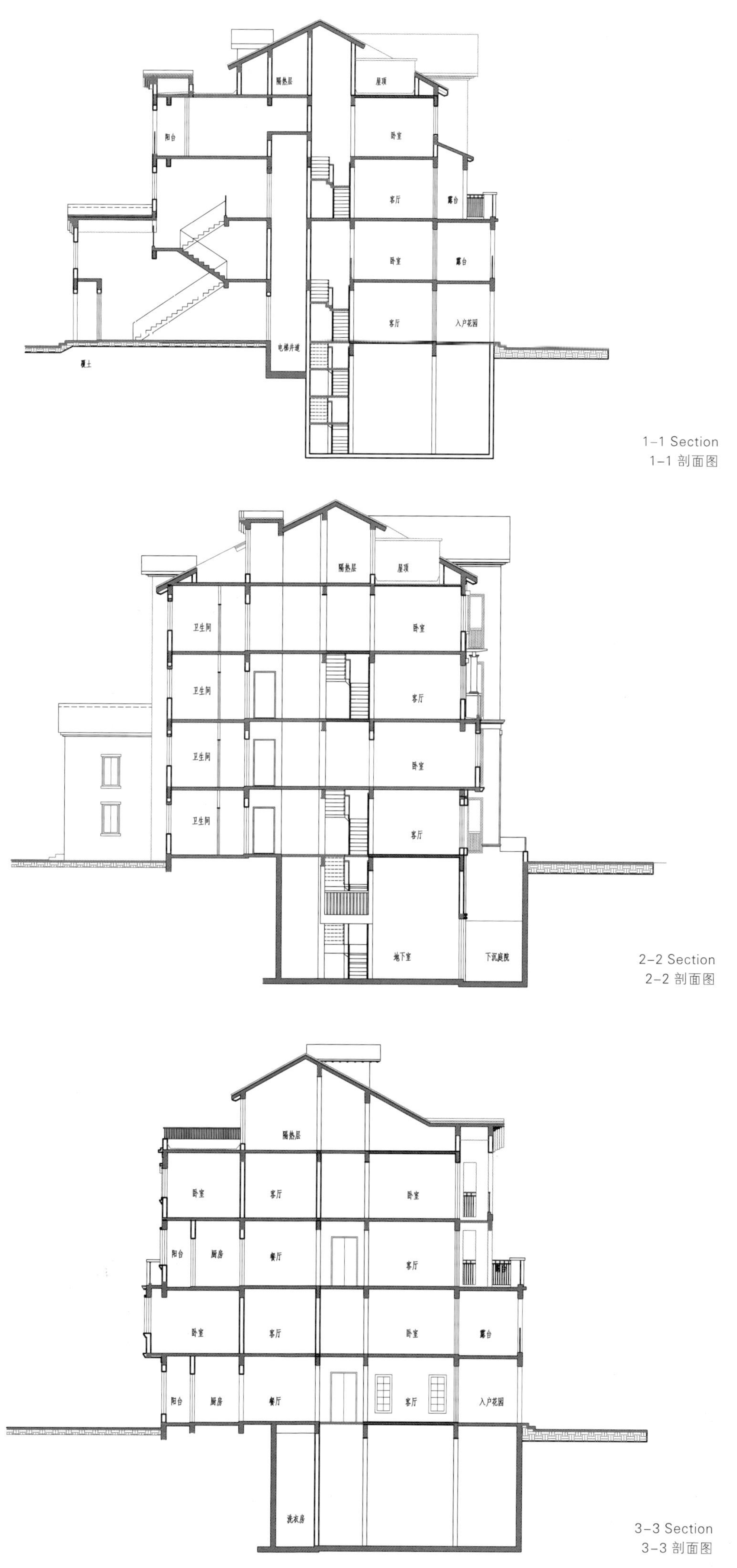

1–1 Section
1–1 剖面图

2–2 Section
2–2 剖面图

3–3 Section
3–3 剖面图

Kunshan Henghai International Golf Holiday Villa (Phase II)

昆山恒海国际高尔夫度假别墅（二期）

Location: Kunshan, Jiangsu, China
Developer: Jiangsu Dongheng Haixin Properties Co., Ltd.
Architectural Design: HIC
Chief Designer: Gao Yingqiang
Land Area: 670,620.6 m^2
Floor Area: 457,964.9 m^2
Plot Ratio: 0.50
Green Coverage Ratio: 45%

项目地点：中国江苏省昆山市
开发商：江苏东恒海鑫置业有限公司
设计单位：上海翰创规划建筑设计有限公司
主创设计：高颖强
占地面积：670 620.6 m^2
建筑面积：457 964.9 m^2
容积率：0.50
绿化覆盖率：45%

KEYWORDS 关键词

Varied House Type
套型多样

Special Style
特殊风貌

Reasonable Layout
布局合理

FEATURES 项目亮点

All buildings take official Spanish style facades, and draw the abundant outline in the horizon through different sizes, heights and the combinations of mutual penetration and backward terraces.

所有建筑采用官式西班牙风格立面，通过建筑体量的大小、体型长短变化或互相穿插的组合以及退台等形式，错落有致地勾勒出丰富的“天际轮廓线”。

Overview

This project is located in Dianshan Lake Tourist Resort (Dianshan Lake waterfront international community) in Kunshan city, with abundant tourism resources. It is adjacent to the Xubao Golf Course with both natural ecological environment and entertainment environment. This planning and design are for phase II which is located at the northeast corner of this site: from Xubao Road on the east, to building lines on the west, from Yongli Road on the north, to the internal landscape riverway on the south.

项目概况

项目地处昆山市淀山湖旅游度假区（淀山湖滨水国际社区）内，旅游休闲资源丰富。紧临旭宝高尔夫球场，自然生态环境与休闲娱乐环境兼具。此次规划设计的项目为二期工程，位于整个地块的东北区：东起旭宝路，西至用地红线，北到永利路，南面以内部景观河道为界。

Architectural Design Principles

1. Varied and flexible house type

Taking the features of traditional Spanish villa and house and combining the demands of contemporary Chinese elites toward modern life, it offers different villa types and house types for different classes.

2. Creating good living environment

Kunshan is cold in winter and hot in summer. To adapt to this climate characteristic, the houses are mainly north-south orientation. The living room, bedroom and kitchen all have natural lighting; the house enjoys north-south air convection with good natural ventilation; most of the living rooms are facing south and at least one bedroom is facing south in every house.

Site Plan
总平面图

3. Reasonable area distribution and indoor space organization

The design makes reasonable area distribution of indoor space, while "making full use of the space" and "making moderate waste". It enlarges the area of living room and the usable area of dining room appropriately, makes comfortable bathroom arrangement and avoids to make hidden bathroom. Every house has at least three bedrooms with an area more than 15 m^2 respectively. Suite room is preferred. The design also enlarges the usable area of the bedroom appropriately and places bathroom, dressing room, study, etc. in the master bedroom.

4. Under the existing area index, the design well organizes and arranges space with complete function, to achieve "separate bedrooms" "separate living room and bedroom" "separate public and private life", forming a part of ritual transition a space based on the area division to show the comfort and luxury of villa life.

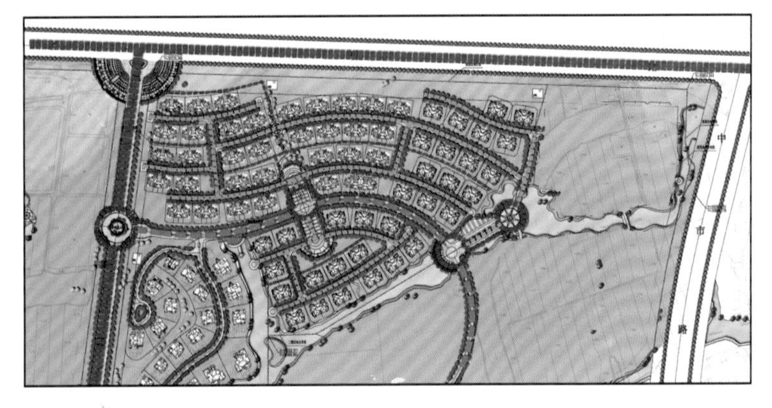

Traffic Drawing
交通分析图

Green Landscape Analysis Drawing
绿化景观分析图

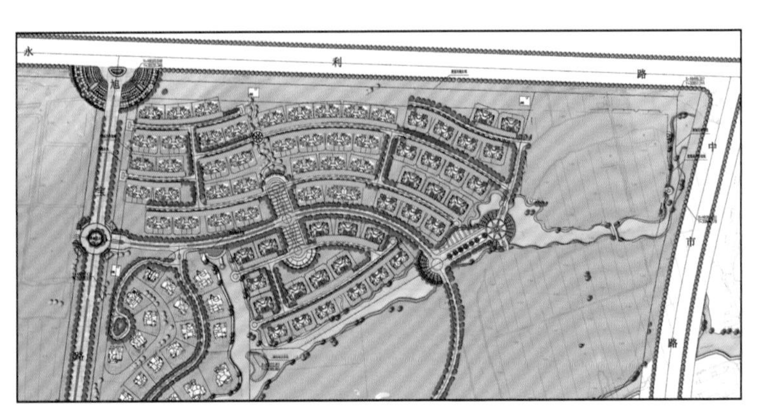

Parking Diagram
停车场分析图

Fire Analysis Diagram
消防分析图

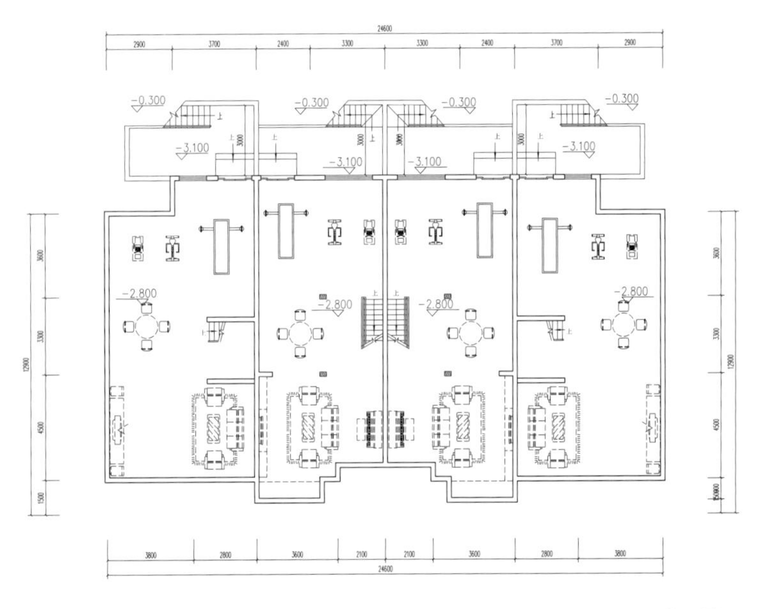

United Villa—Plan for Basement One Floor
联排别墅—地下一层平面图

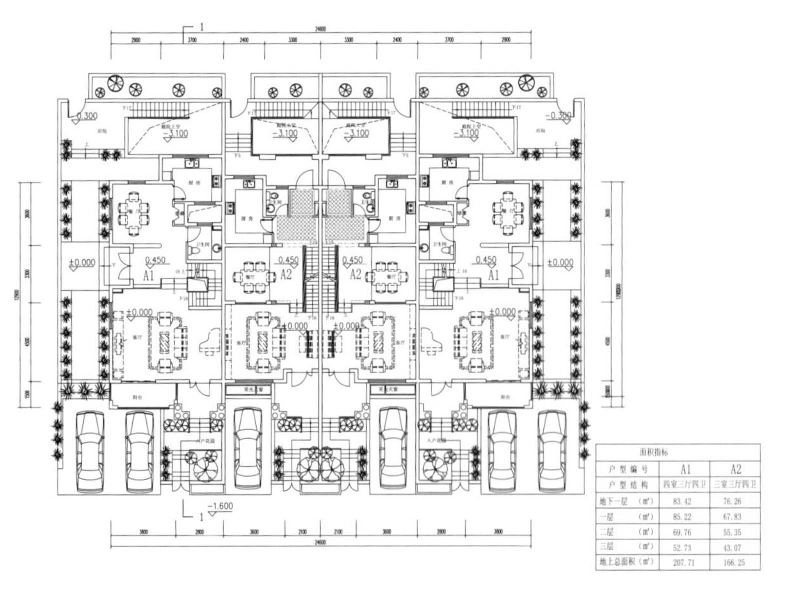

面积指标		
户型编号	A1	A2
户型结构	四室三厅四卫	三室三厅四卫
地下一层 （m²）	83.42	76.26
一层 （m²）	85.22	67.83
二层 （m²）	69.76	55.35
三层 （m²）	52.73	43.07
地上总面积（m²）	207.71	166.25

United Villa—First Floor Plan
联排别墅——层平面图

➤ 建筑设计原则

1. 套型灵活多样

吸收传统西班牙别墅民居的特点，结合当代中国精英阶层对现代生活的需求，提供适应不同阶层需要的别墅类房型。

2. 创造良好的居住环境

昆山冬季寒冷、夏季炎热，为适应当地的气候特点，住宅朝向以南北向为主。起居室、卧室及厨房均有自然采光，房内空气南北对流，自然通风良好。起居室以朝南为主，并且每户至少有一个卧室朝南。

3. 合理分配面积和组织户内空间

合理分配户内各功能空间的面积，做到“大而有当”“适度浪费”。适当扩大起居室的面积和餐厅的使用面积，卫生间分室布置体现舒适性，且尽量为明厕。每户至少拥有3个15 m^2以上的卧室，房间尽量做成套间，适当扩大卧室的使用面积，主卧套型设置卫生间、更衣间和书房等。

4. 在现有的面积指标下，做到功能齐全、合理组织和布置各功能行为空间，达到“寝分离”“居寝分离”“公私分离”，在动静分区的基础上形成部分仪式化的过渡空间，体现别墅生活的舒适性、豪华性。

United Villa—Second Floor Plan
联排别墅—二层平面图

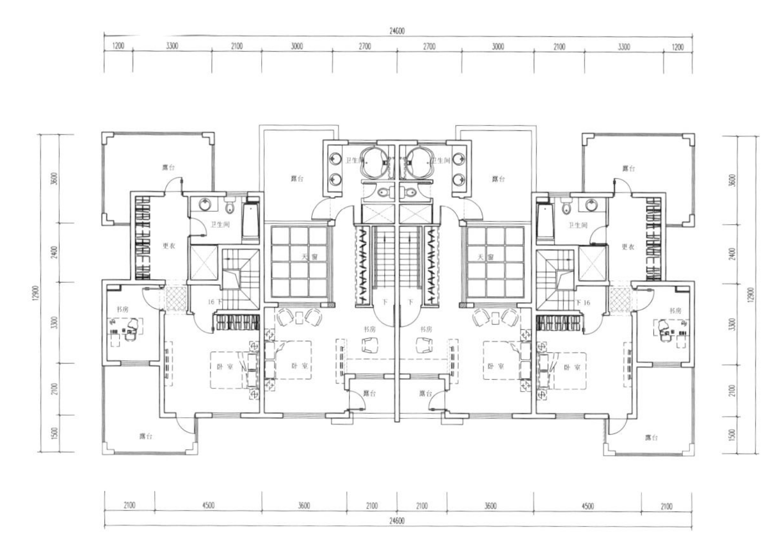

United Villa—Third Floor Plan
联排别墅—三层平面图

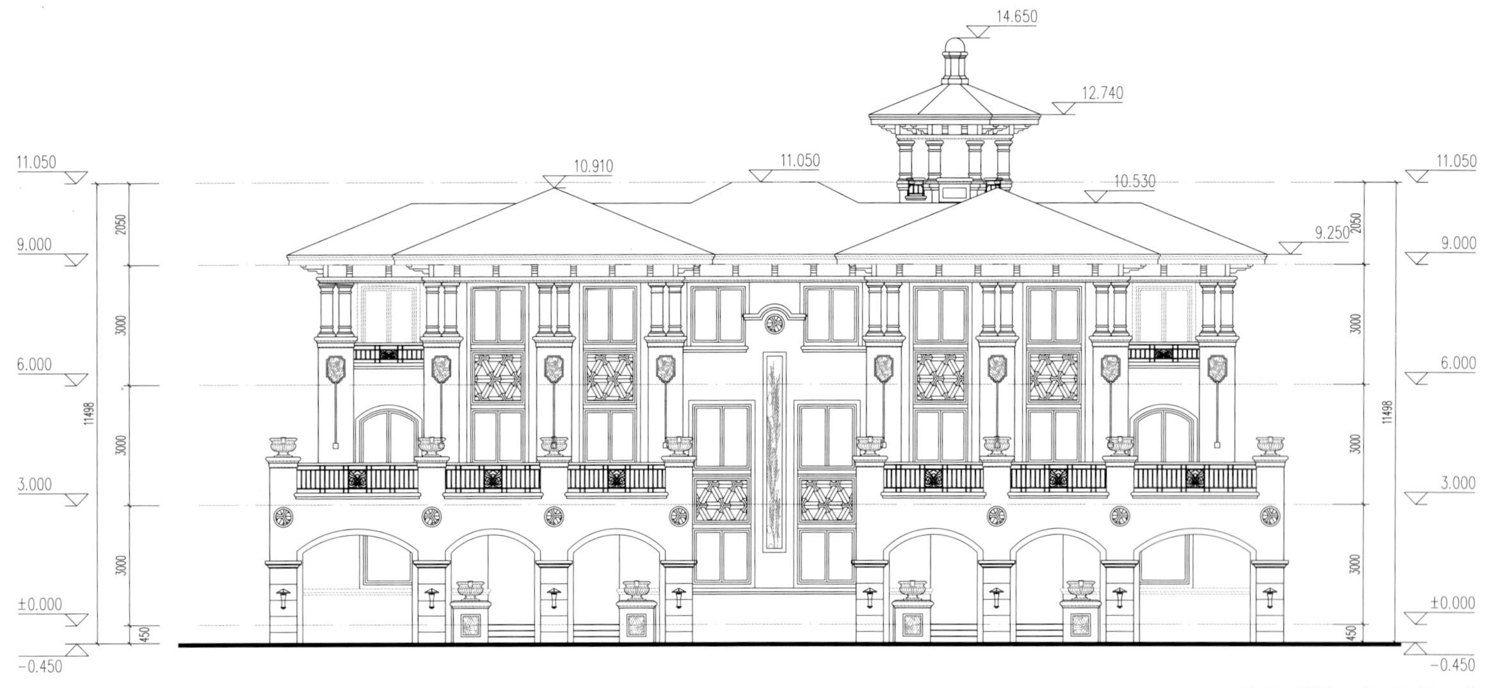

United Villa—North Elevation
联排别墅—北立面图

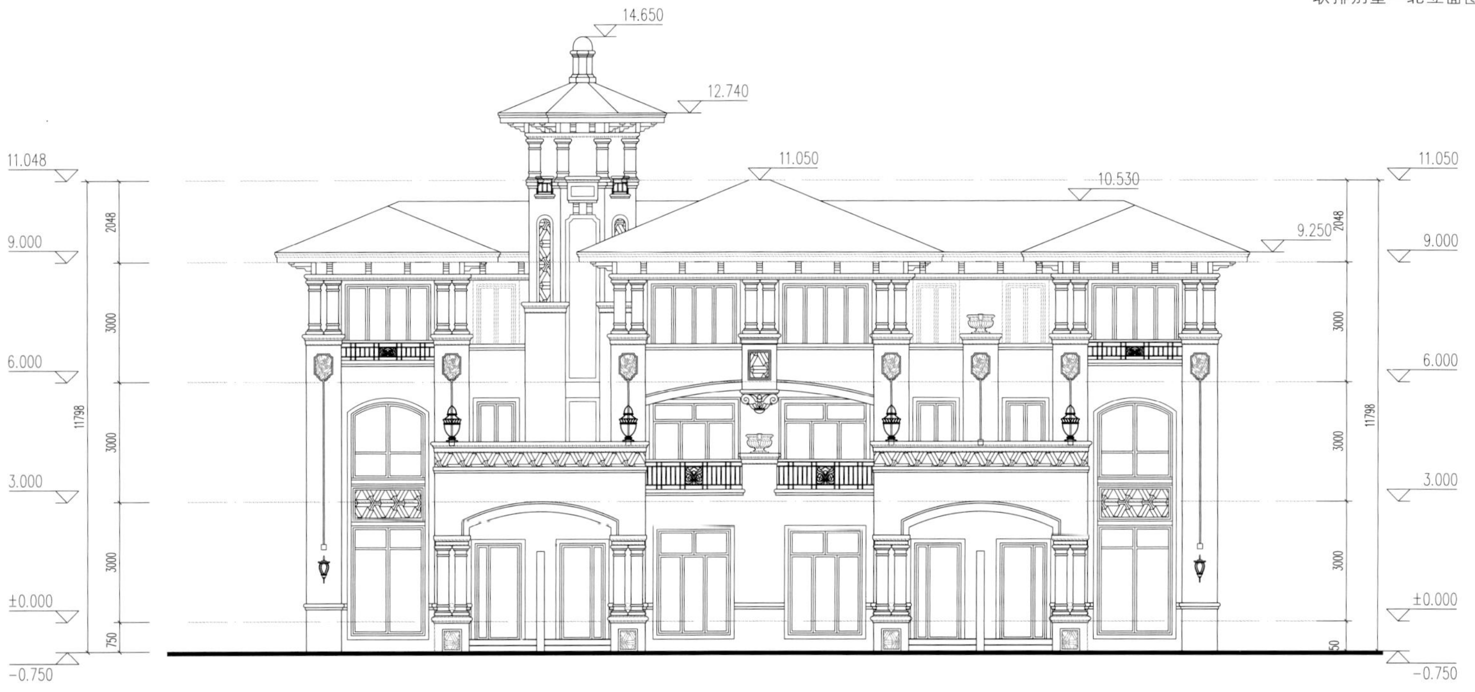

United Villa—South Elevation
联排别墅—南立面图

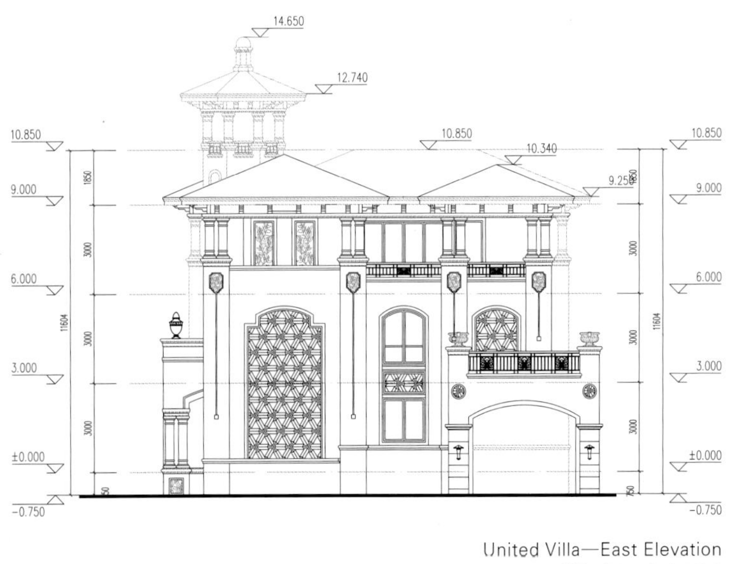

United Villa—East Elevation
联排别墅—东立面图

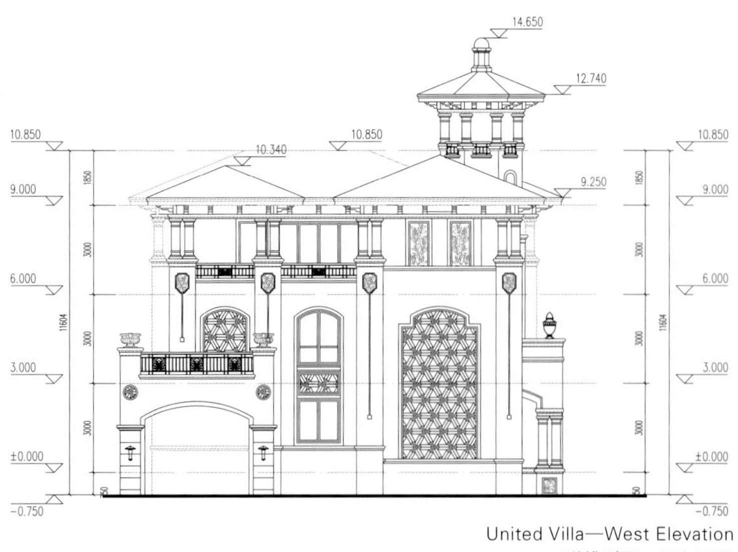

United Villa—West Elevation
联排别墅　西立面图

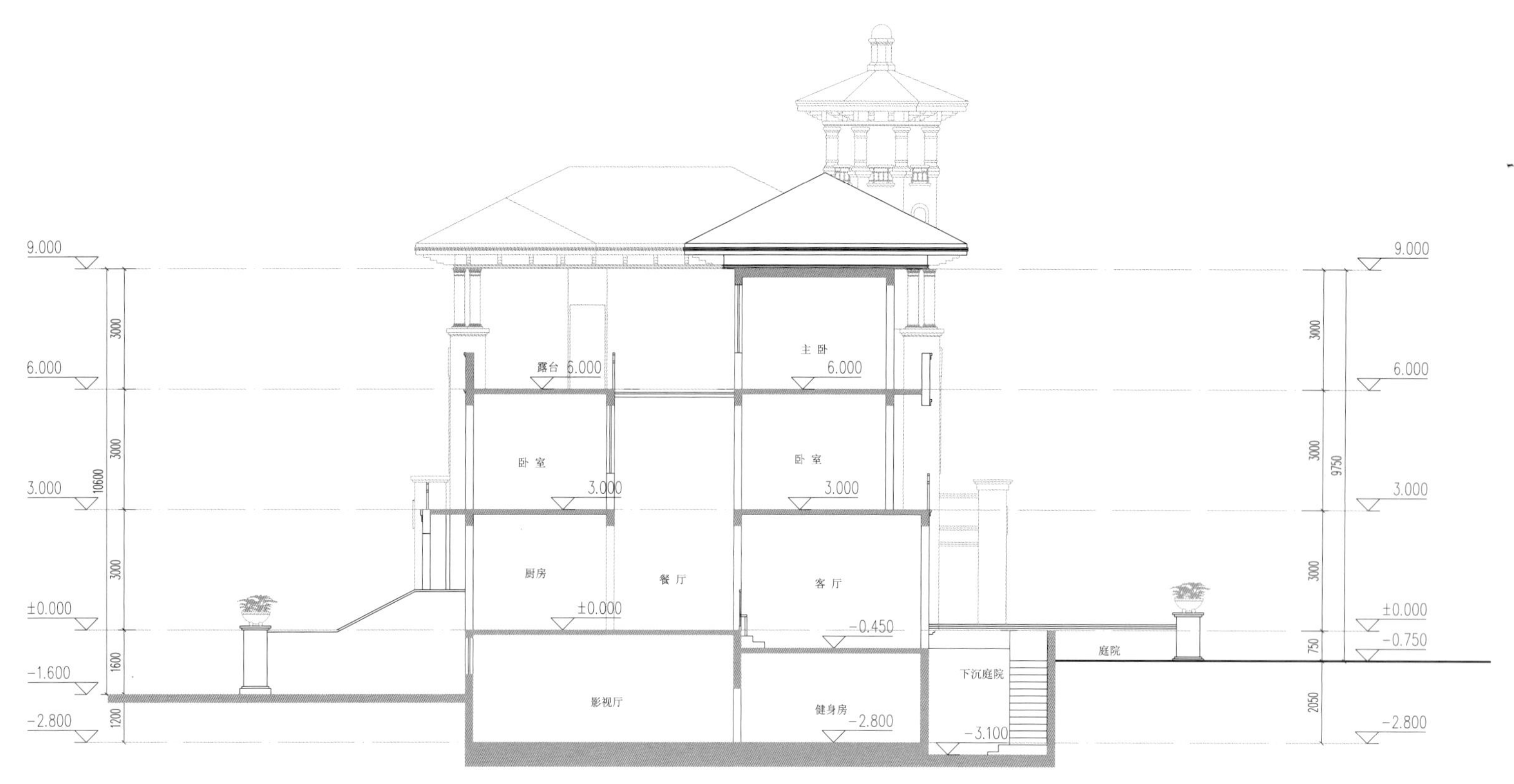

United Villa 1-1 Section
联排别墅 1-1 剖面图

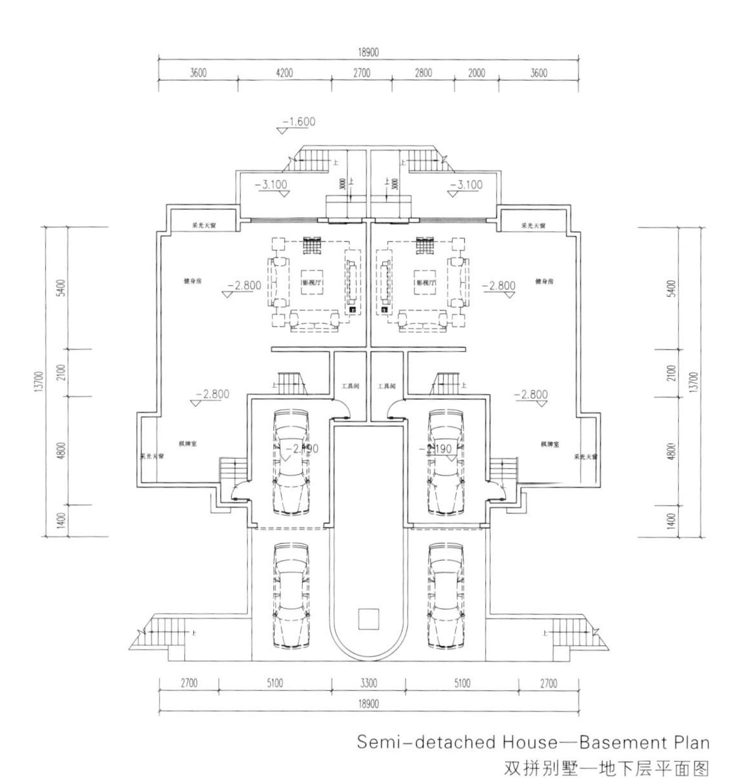

Semi-detached House—Basement Plan
双拼别墅—地下层平面图

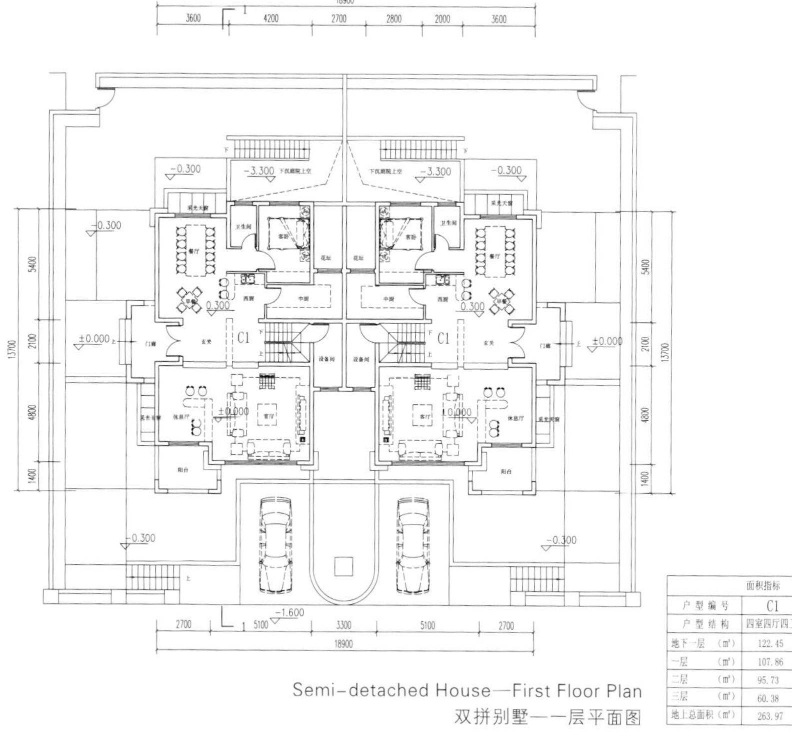

Semi-detached House—First Floor Plan
双拼别墅—一层平面图

面积指标	
户 型 编 号	C1
户 型 结 构	四室四厅四卫
地下一层 （㎡）	122.45
一层 （㎡）	107.86
二层 （㎡）	95.73
三层 （㎡）	60.38
地上总面积（㎡）	263.97

Semi-detached House—Second Floor Plan
双拼别墅—二层平面图

Semi-detached House—Third Floor Plan
双拼别墅—三层平面图

Semi-detached House—South Elevation
双拼别墅—南立面图

Semi-detached House—North Elevation
双拼别墅—北立面图

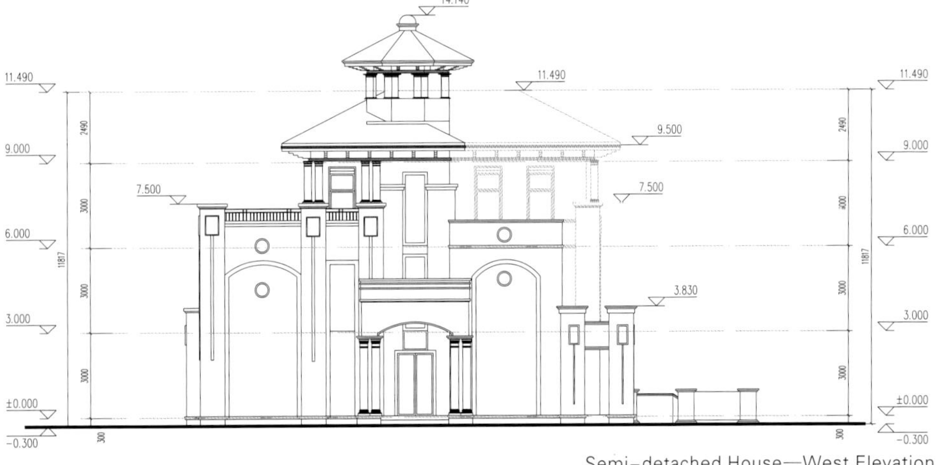

Semi-detached House—West Elevation
双拼别墅—西立面图

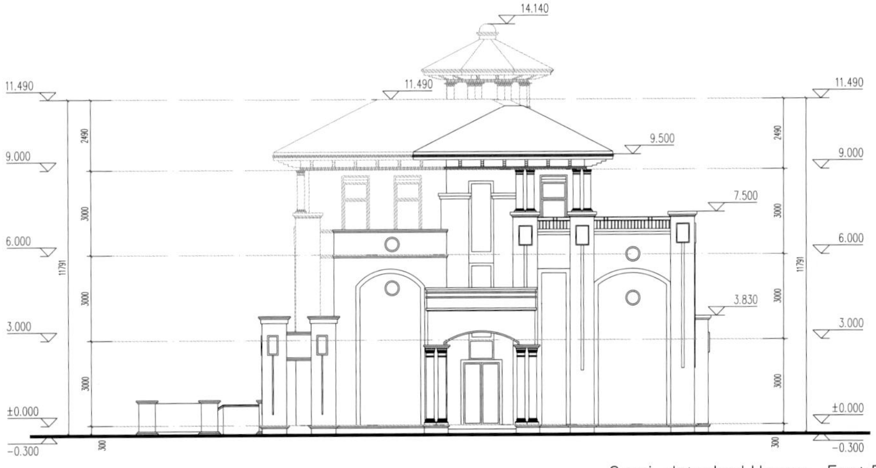

Semi-detached House—East Elevation
双拼别墅—东立面图

➤ Architectural Image Design

All buildings take official Spanish style facades, and draw the abundant outline in the horizon through different sizes, heights and the combinations of mutual penetration and backward terraces.

The design makes artistic treatment to the balcony, handrail and other parts to make it form an architecture symbol with special features and characteristics, and repeat them to stress on the integrity of the residential community and produce the sense of rhythm. At the same time, it focus on details processing of cornice, frieze, etc. to enrich the whole and make people feel at home.

The outer wall takes high-grade stone with warm color and light yellow brown coating paint to make the building elegant and the living area peaceful and pleasant.

➤ 建筑造型设计

所有建筑采用官式西班牙风格立面，通过建筑体量的大小、体型长短变化或互相穿插的组合以及退台等形式，错落有致地勾勒出丰富的“天际轮廓线”。

对阳台、栏杆等部分进行艺术处理，使其形成具有特殊风貌的建筑符号。反复使用，强调住宅群体的整体性，并赋予其韵律感。同时重点处理檐口、腰线等细部，丰富整体，使人倍感亲切。

外墙面以暖色高档石材加浅黄褐的涂料为主，既使建筑物典雅，又使居住区幽静宜人。

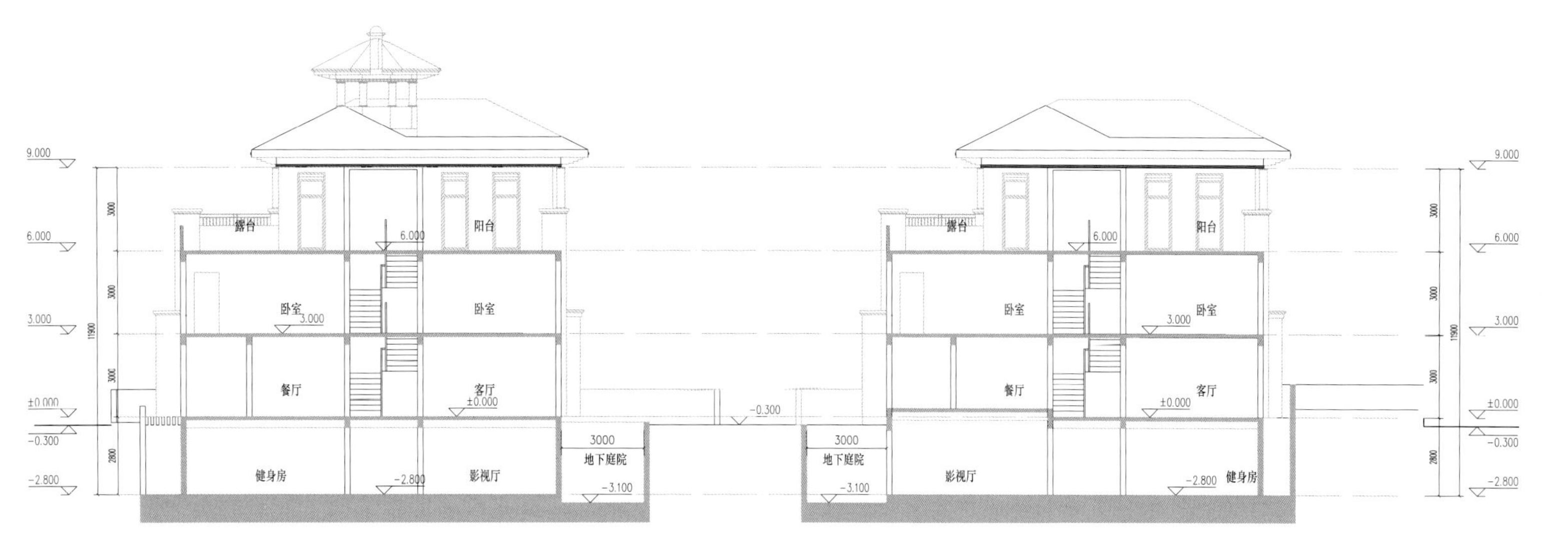

Semi-detached House 1-1 Section
双拼别墅 1-1 剖面图

Hidden Valley, Hangzhou

杭州莱蒙水榭山别墅

Location: Hangzhou, Zhejiang, China
Developer: Top Spring Real Estate (Fuyang) Company Limited
Architectural Design: Shenzhen Huahui Design Co., Ltd.
Landscape Design: Bensley Design Group International Consultants Co., Ltd.
Land Area: 280,000 m^2
Floor Area: 344,600 m^2
Plot Ratio: 1.10
Green Coverage Ratio: 30%
Date of Design: April, 2010
Status: Phase Ⅰ was completed in June, 2011; phase Ⅱ is under construction.

项目地点：中国浙江省杭州市
开发商：莱蒙置业（富阳）有限公司
建筑设计：深圳市华汇设计有限公司
景观设计：Bensley设计集团国际咨询有限公司
占地面积：280 000 m^2
建筑面积：344 600 m^2
容积率：1.10
绿化覆盖率：30%
设计时间：2010年4月
项目状态：2011年6月一期完成，二期在建

KEYWORDS 关键词

Southeast Asian Style
东南亚风情

Big Slope Roof
大坡屋顶

Enclosed Courtyard
围合院落

FEATURES 项目亮点

Integrated with the cultural context of Jiangsu and Zhejiang regions, the facade design uses the elements and symbols in Southeast Asian buildings to present large slope rooves and bright colors. In addition with the love to stone materials, it creates a kind of stylish and quality villa style.

建筑立面融合江浙文脉，从东南亚建筑中提取元素符号，以挑檐较深的大坡屋顶与明快的外观色彩，结合杭州别墅市场对石材的要求，创造出风情感与品质感兼具的别墅风格。

➤ Project Orientation

The Hidden Valley, Hangzhou is planned as a development for low-density townhouses and courtyard villas which are connected by a walkway running through the Valley. Taking advantages of the superior natural environment, the houses and villas in Southeast Asian style are built along the north branch of the river, which enables people to live in the courtyard along the street, or live on the island, or live in the valley and mountains. Developed with the idea of "landscape residence and resort style", the Hidden Valley is envisioned to be a supersized high-end community which integrates high-rise apartments, low-density townhouses and luxurious single-family villas.

➤ 项目定位

莱蒙水榭山规划为低密度排屋、家族院墅等产品。规划中设计了一条谷居人行流线，串联别墅区不同类型的产品。以东南亚情景院墅为载体，依北支江蜿蜒之势顺势而生，规划以街居、山居、岛居、谷居和院居五重形态为主。项目以山水人居、度假风情为开发理念，规划为集高层景观公寓、低密度排屋及超高端独栋别墅于一体的超大型高端社区。

➤ Integrated Architectural Style

The design has integrated the architectural style in Southeast Asia with the climate and culture of Jiangsu and Zhejiang regions. Big slope roof, stone and metal surface form the building facades which cater to the real estate market of Hangzhou City. It has also used the proportion, composition as well as the typical elements and symbols in the Southeast Asian buildings, to highlight the quality of the project and respond to the project orientation.

➤ 建筑风格融合

项目萃取现代东南亚建筑文化精髓，融合江浙的气候及人文气质，将两者自然巧妙地运用到建筑设计当中。立面以挑檐较深的大坡屋顶、石材及金属墙面，迎合杭州市场对别墅品质的要求，并从东南亚建筑中提取了一些比例、构图及元素符号呼应整个项目定位。

➤ Architectural Facade Design

Integrated with the cultural context of Jiangsu and Zhejiang regions, the facade design uses the elements and symbols in Southeast Asian buildings to present large slope rooves and bright colors. In addition with the love to stone materials, it creates a kind of stylish and quality villa style. Granite cladding is skillfully contracted to form a heavy building volume, while the horizontal lines on the facade will alleviate this kind of heaviness. Moreover, elegant blocks combine with the Southeast Asian style capmould and windows to create a luxurious and comfortable living atmosphere.

➤ 建筑外立面设计

建筑立面融合江浙文脉，从东南亚建筑中提取元素符号，以挑檐较深的大坡屋顶与明快的外观色彩，结合杭州别墅市场对石材的要求，创造出风情感与品质感兼具的别墅风格。建筑以干挂石材做微妙的收分，形成厚重敦实的建筑体量，通过对立面横向线条的加强，削弱了建筑的体量感。同时用简洁的虚实体块对比及一些东南亚风情特有的压顶线脚、窗户手边等细节，营造出尊贵且舒适的居住氛围。

Site Plan
总平面图

➤ Enclosed Courtyard

Inspired by the image of Monet's "Waterlilies", different residential products for the "island life" are built in group to form a series of enclosed courtyards for neighborhood communications. Amid these courtyard spaces, large landscape node is created to provide great views. This kind of layout is cohesive to promote the possibilities of communications between neighbors. In addition, the courtyards combine with the water features to provide water views and keep privacy at the same time.

➤ 围合院落

最具特色的"岛居生活"组团源自莫奈"睡莲"的意象，以不同产品围合成一个个院落，院落中形成公共的邻里交往空间，院落与院落之间亦围合成大的景观节点。合院产品具有很强的向心力，增强了邻里之间交往的可能性。同时合院与水景结合，户户临水，景观与私密性俱佳。

➤ Multiple Private Spaces

Based on the Southeast Asian leisure style, it uses such traditional landscape skills as appositive scenery and borrowed scenery to create multiple courtyard spaces. By installing large-area glass doors or windows, it has introduced the courtyard views into the dining room and living room. In this way, the interior spaces are well connected with the exterior spaces. Meanwhile, green landscapes in the side yard and small patio will provide the walkways, bathrooms and other functional spaces with unique views. And the private sunken courtyard not only forms multi-level landscapes but also brings abundant daylight and natural wind down to the basement floor.

➤ 多重私密空间

平面以东南亚休闲风情为主题，讲究借景对景的手法，营造多重的院落空间。餐厅、客厅所形成的室内大空间通过大片的玻璃门窗向私家院落敞开，使院落的景观得到完全的渗透，达到室内外空间的融合。同时着力营造一些侧院及小天井的空间绿化景观，使走道与卫生间等一些次要功能房间亦能达到“移步换景”的景观效果。而极具私密感的下沉庭院不仅丰富了院落景观层次，也为地下室提供了充足的采光和通风。

Huizhou Minmetals Hallstatt, Phase I, Area 4 & 5

惠州五矿哈施塔特一期四区、一期五区

Location: Huizhou, Guangdong, China
Developer: China Minmetals Corporation Huizhou Branch
Architectural Design: Pofart Architecture Design Company Limited
Gross Land Area: 31,637 m^2
Gross Floor Area: 21,466 m^2
Plot Ratio: 0.68
Green Coverage Ratio: 40%

项目地点：中国广东省惠州市
开发商：五矿建设惠州公司
建筑设计：深圳市华域普风设计有限公司
总占地面积：31 637 m^2
总建筑面积：21 446 m^2
容积率：0.68
绿化覆盖率：40%

KEYWORDS 关键词

Visual Impact
视觉冲击

Staggered Method
错叠手法

Austrian Style
奥地利风情

FEATURES 项目亮点

The architectural facade reproduces the authentic architectural features of the Austrian Hallstatt town through the study of the details of building such as volume, roof, architrave, flower bed and handrail.

建筑外立面通过对建筑的体量、屋顶、线脚、花池、栏杆等细节的推敲，原汁原味的还原出奥地利哈施塔特风情小镇的建筑风貌。

➤ Overview

Huizhou Minmetals Hallstatt, Phase I, Area 4 & 5 is located in Boluo town of Huizhou city, and the topography of the plot is a sloping lakeside, which is in front of the mountain and faces the lake with the highest altitude difference of 40m. The north side of the plot is the Hallstatt landscape area with purple flowers and streams as the main landscape node, and the west side faces the lake landscape area and the town, while the east and south sides are near the luxury residential area of phase I.

➤ 项目概况

惠州五矿哈施塔特一期四区，一期五区项目位于惠州市博罗县，地块属于滨湖坡地类型，背山面湖，最大高差约40 m。地块北侧为哈施塔特风景区内主要景观节点紫花溪岸，西侧面向湖景及小镇，东侧、南侧临一期别墅住宅区。

➤ Overall Planning

The design of the overall planning strives to maximally use the altitude difference of the topography and the landscape resources such as the lake and the town and adjusts to the local conditions to incisively and vividly manifest the features of the sloping lakeside, effectively control earth excavation and the height of the barricade, and avoid unnecessary construction cost.

➤ 整体规划

整体规划设计上力求做到最大限度的利用地形高差及湖景小镇景观资源，依山就势，将山地滨湖住宅的特色发挥得淋漓尽致，有效的控制了土方开挖和挡墙高度，避免没有必要的施工浪费。

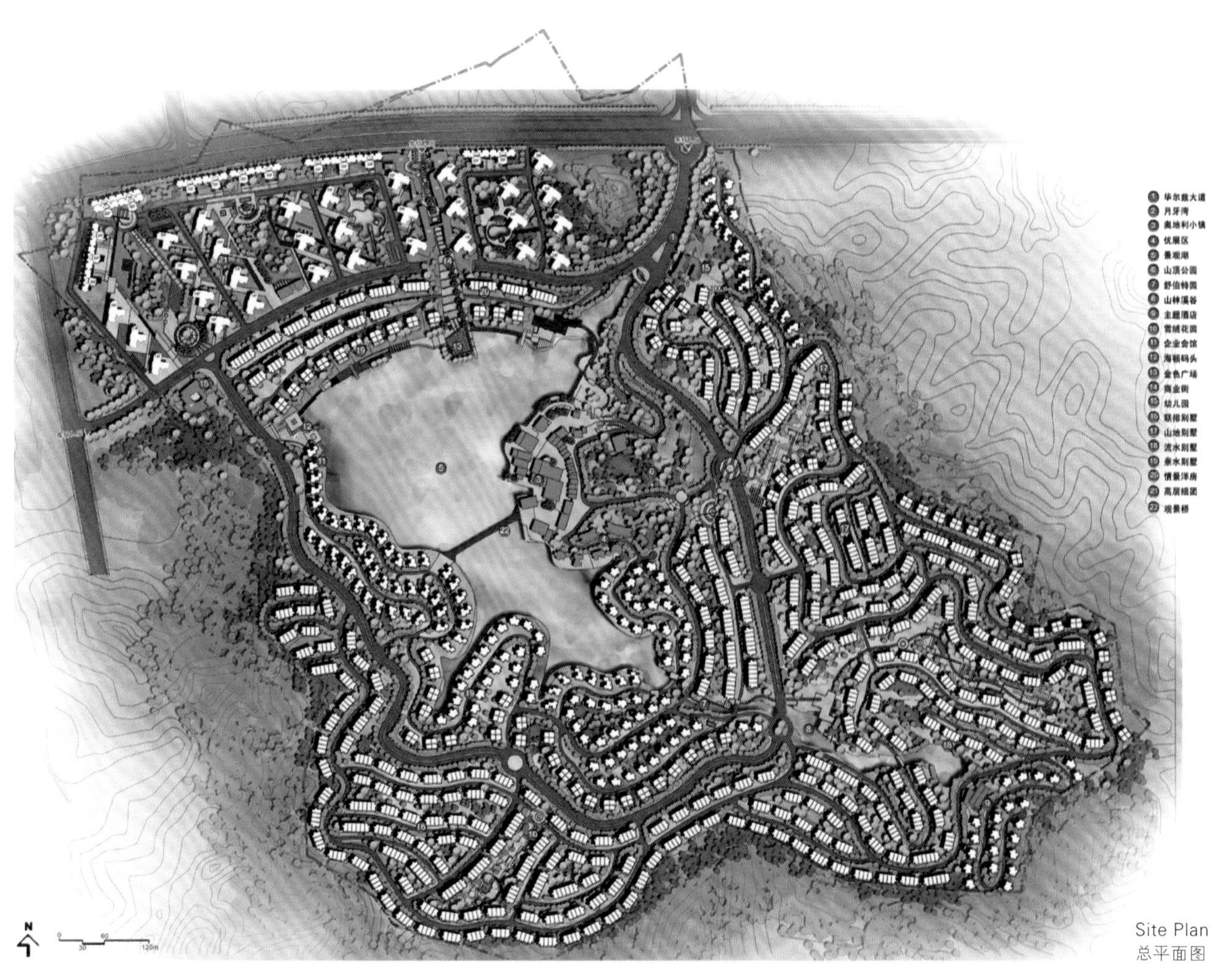

Site Plan
总平面图

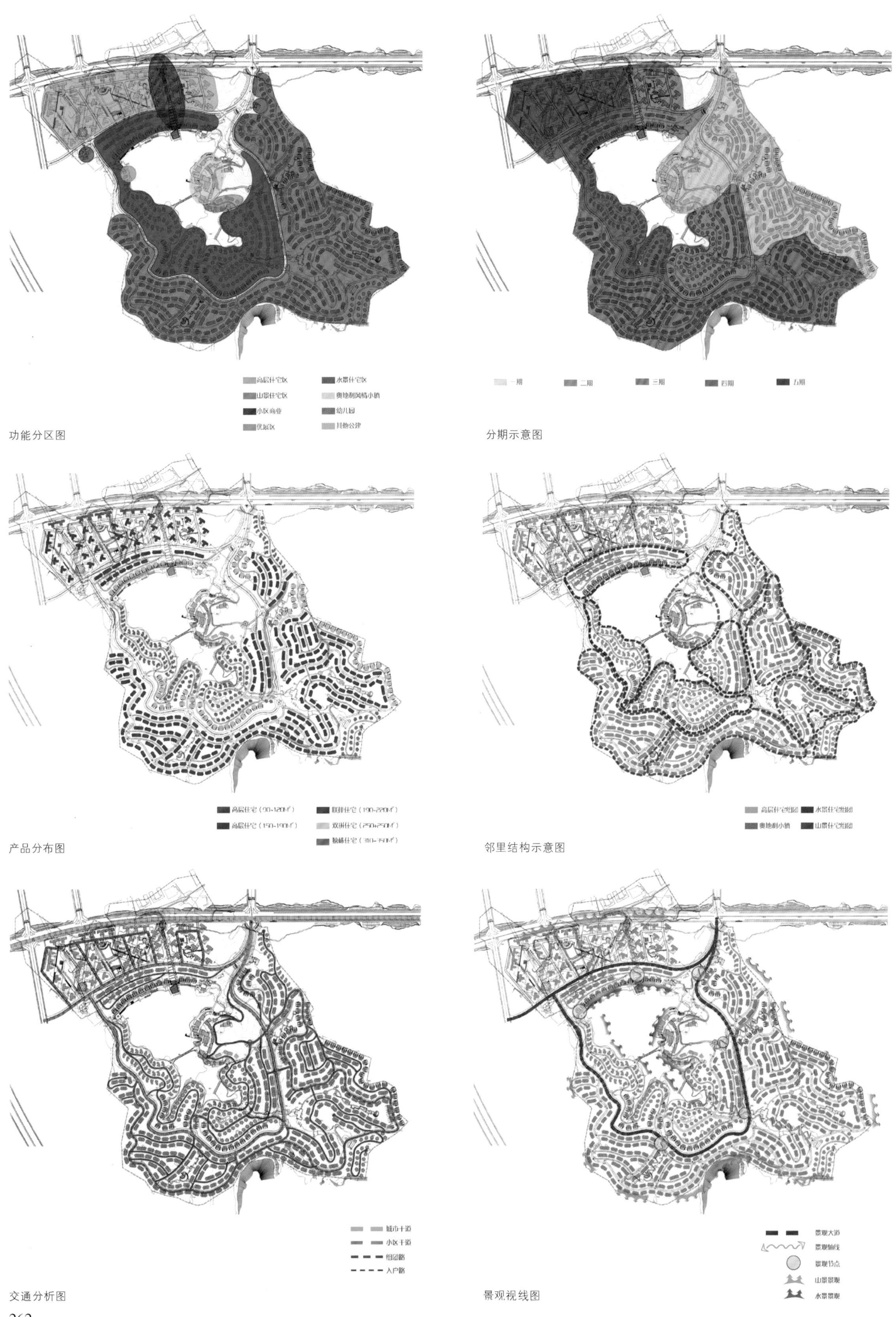

功能分区图

分期示意图

产品分布图

邻里结构示意图

交通分析图

景观视线图

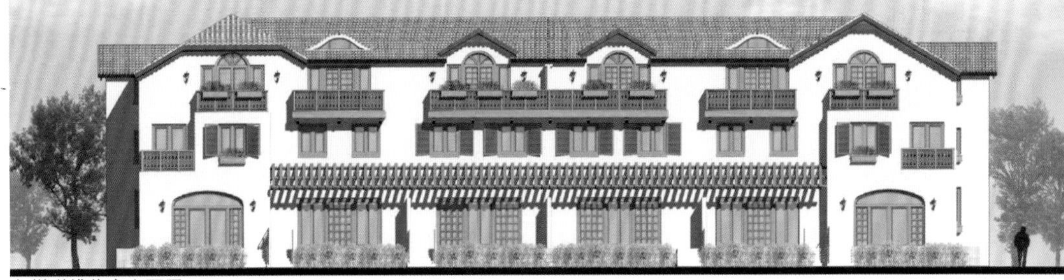
六拼联排住宅立面图

独栋住宅立面图

四拼联排住宅立面图

叠拼住宅立面图

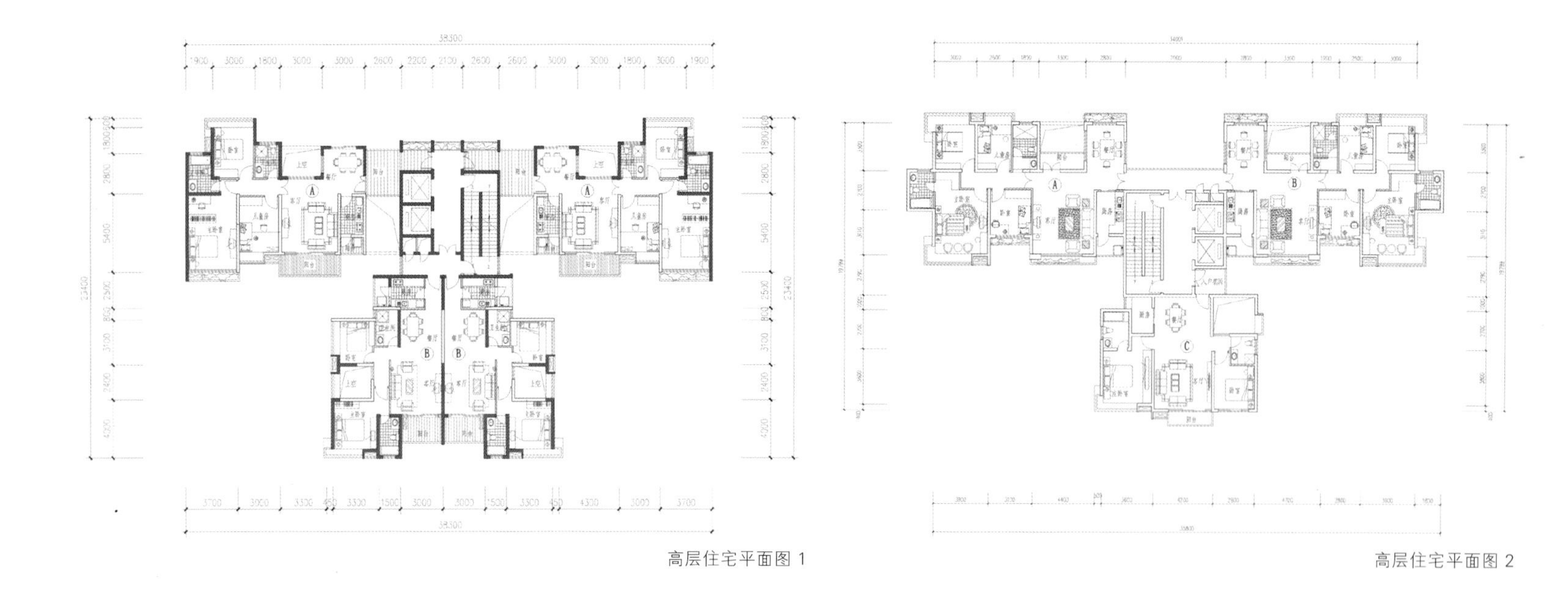

高层住宅平面图 1

高层住宅平面图 2

➤ Design Idea

The project uses the Austrian Hallstatt town as design blueprint, and with the highest gradient of near 45 degrees, it adjusts measures to the local conditions and uses the least volume of earth to resolve the altitude difference of the giant site. The project makes full use of the relationship between the lake landscape and the site and combines the altitude difference of the terrain to creatively design and plan orderly staggered Austrian community with beautiful environment.

➤ 设计理念

项目以奥地利风情小镇哈施塔特为设计蓝本，在最陡坡度接近45度的山地条件下，因地制宜，以最少土方量化解巨大场地高差。项目充分利用湖景及场地关系，结合地形高差创新设计，规划出错落有致、环境优美的奥地利风情社区。

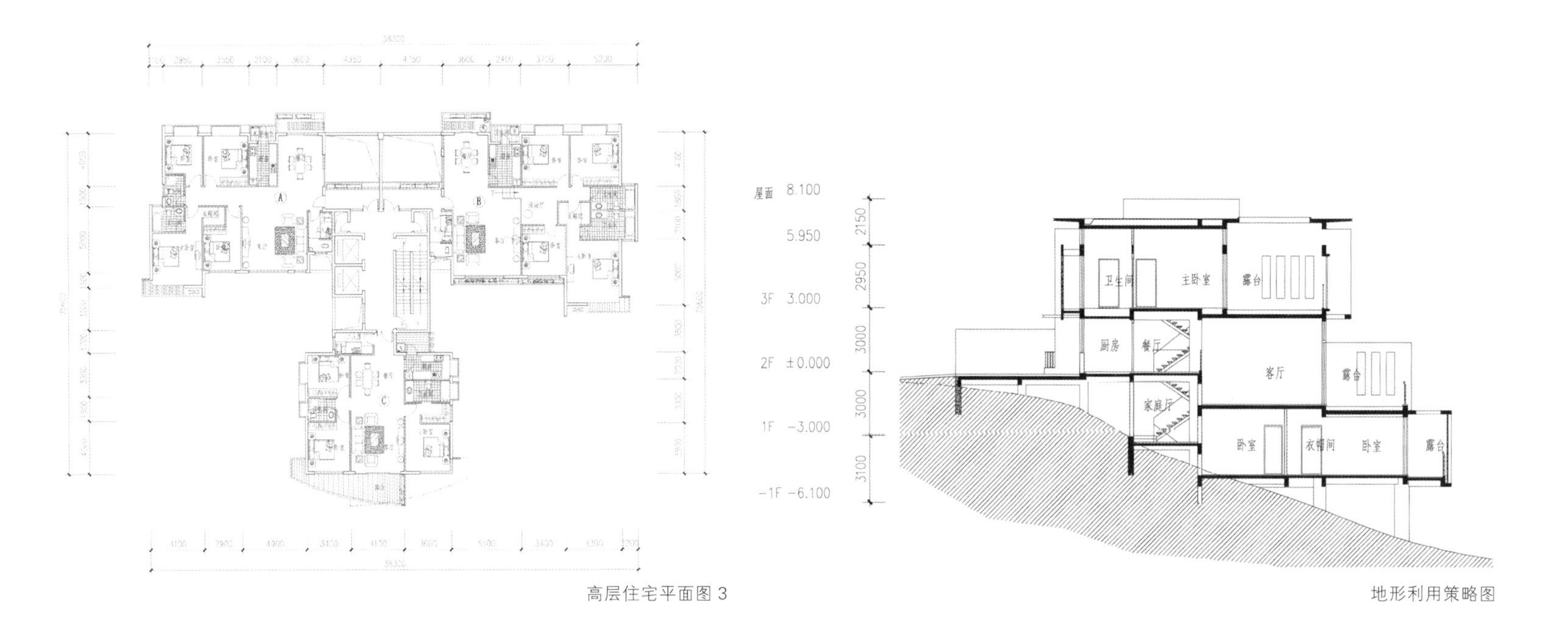

高层住宅平面图 3

地形利用策略图

➤ Architectural Design

The architectural facade reproduces the authentic architectural features of the Austrian Hallstatt town through the study of the details of building such as volume,roof, architrave, flower bed and handrail. The small high-rise apartment of the fifth area, Phase I fully derives the outline of mountain and texture of water to create a featured residential building with a strong visual impact in this landscape area.

➤ 建筑设计

建筑外立面通过对建筑的体量、屋顶、线脚、花池、栏杆等细节的推敲，原汁原味的还原出奥地利哈施塔特风情小镇的建筑风貌；一期五区小高层公寓更是汲取山的轮廓和水的肌理，打造出一个具有视觉冲击力的风景区特色住宅建筑。

➤ House Type Design

The plan design of monomer building is mainly in small-type apartment, and it uses staggered methods layer by layer to ensure each household with a lake view terrace. It also designs private gardens at the entrance floor by using the angle of architectural layout to promote the comfort and quality of each household.

➤ 户型设计

单体平面设计以小户型公寓为主，采用层层错叠手法，保证每户有观湖大露台，并利用建筑布局的角度，在入户层设计私家花园，提升每户的居住舒适度和品质感。

独栋住宅平面图 一层平面图

独栋住宅平面图 二层平面图

独栋住宅平面图 三层平面图

独栋住宅平面图 三层平面图

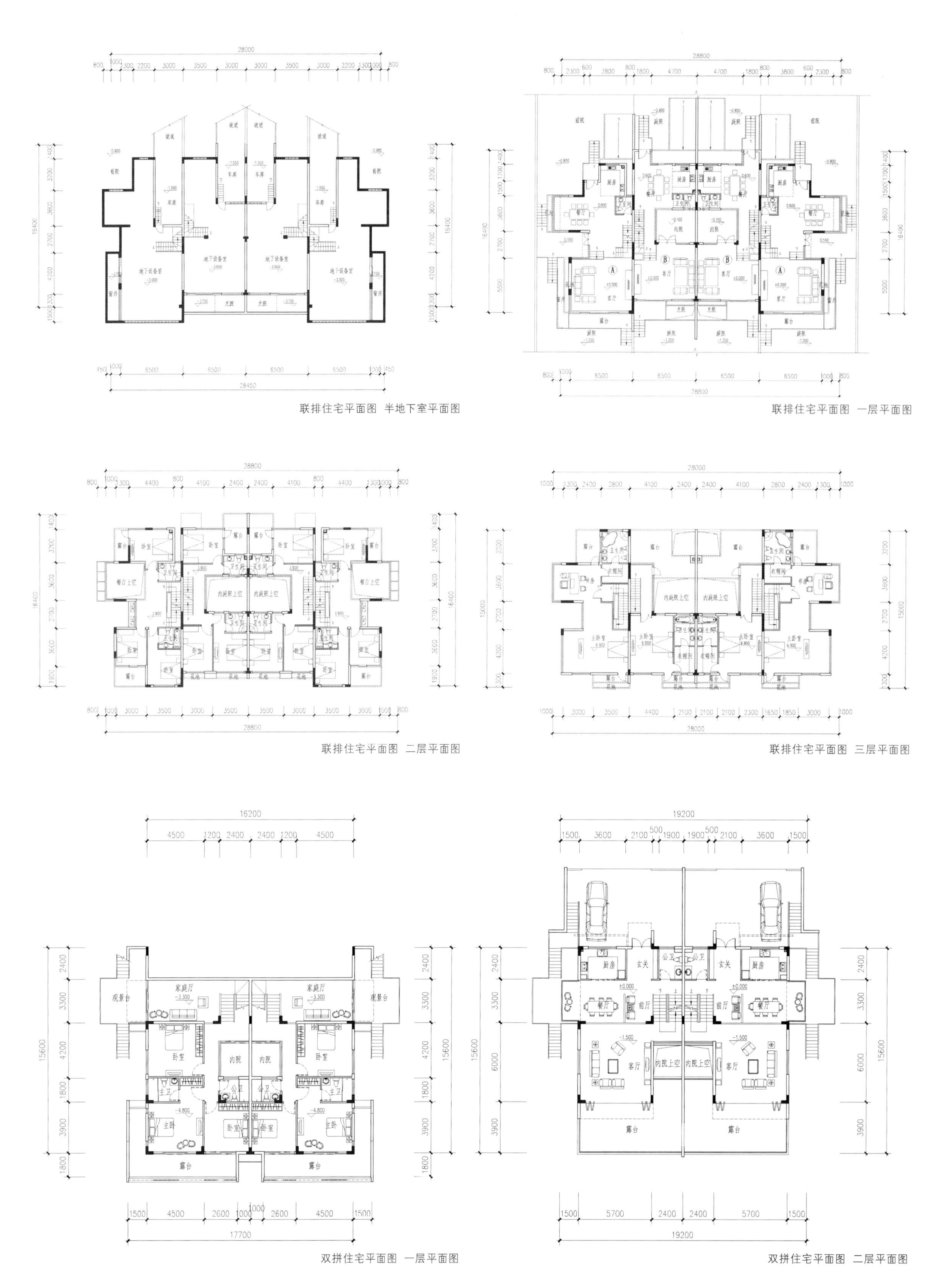

联排住宅平面图 半地下室平面图

联排住宅平面图 一层平面图

联排住宅平面图 二层平面图

联排住宅平面图 三层平面图

双拼住宅平面图 一层平面图

双拼住宅平面图 二层平面图

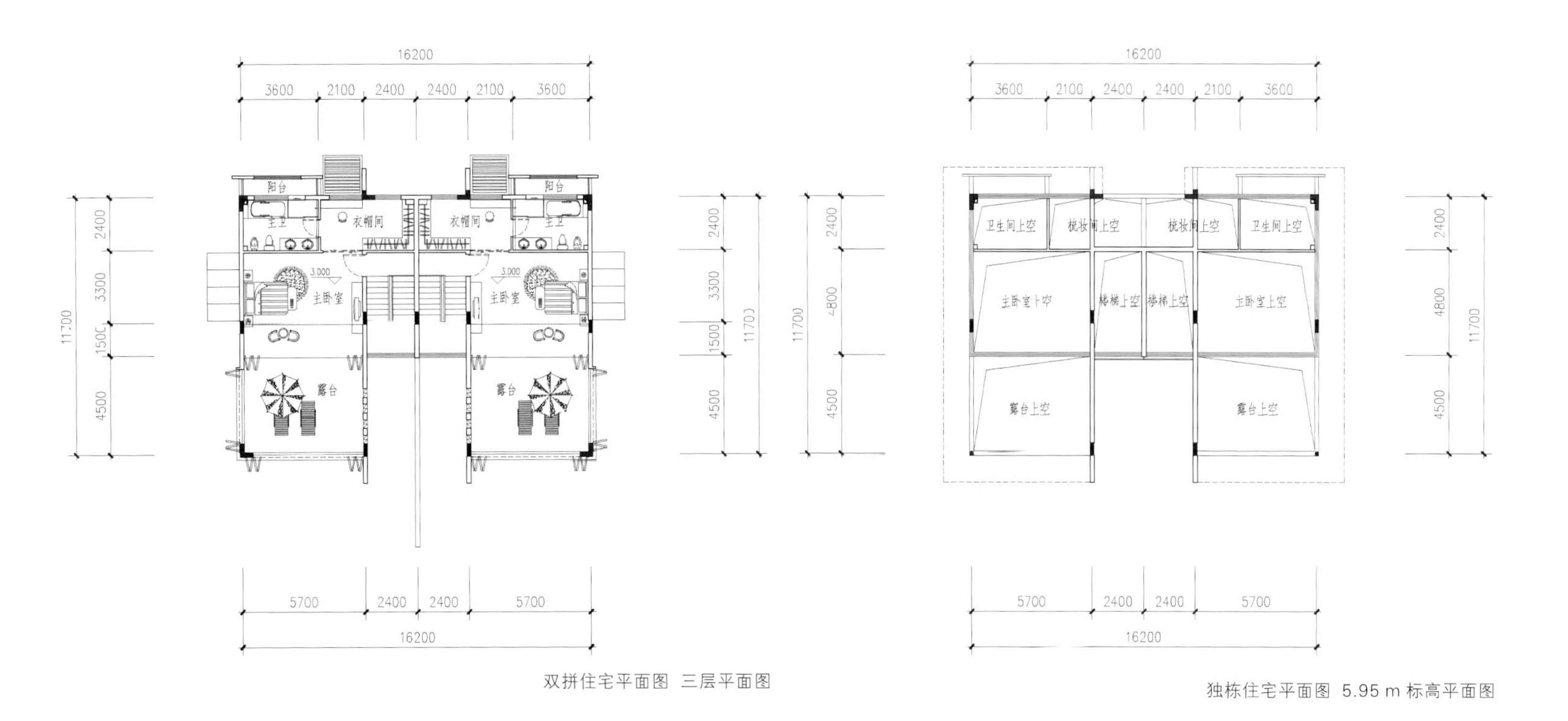

双拼住宅平面图 三层平面图

独栋住宅平面图 5.95 m 标高平面图